Whole Wheat Cookery

TREASURES FROM THE WHEAT BIN

Howard and Anna Ruth Beck

Whole Wheat Cookery

Library of Congress
Catalog Card Number 86-72596

ISBN 1-1882420-00-4
(Previous ISBN 0-9617875-0-3)

First Edition, 1993
Hearth Publishing, Hillsboro, Kansas

Edited and compiled by Howard and Anna Ruth Beck
Designed by Ken D. Gingerich

Printed in the United States of America
by Multi Business Press

FOREWORD

Native Americans and early pioneers may not have fully understood the nutritional value of whole grain. Most importantly to them, it was edible. The fact that whole grain could be ground into flour between two rocks made it a versatile ingredient, and it must have been more palatable than munching whole kernels.

Eventually, settlers planted tillable crops like wheat and milo. They also developed the technology to manage the crops. These farmers soon built wind driven mills to grind whole wheat into flour.

As farming grew in the midwest, vast tracts of fertile prairie grassland gave way to the plow. What was once considered a formidable part of the "Great American Desert" became the "Bread Basket of the World." This arid environment produced so much grain that home storage and wind driven flour milling became impractical.

Before long, huge cement structures began to rise above the golden ocean of wheat waving in the wind. These elevators were grain storage units with attached milling operations, and most were formed as cooperatives by area farmers.

Today, nearly every farming community in Kansas has at least one of these "Prairie Skyscrapers." Stark white, these towering pinnacles can be seen for miles and embody the very imagery of bread itself—freshly baked loaves standing side-by-side on end.

Wheat has always been at the heart of bread baking. Although it has great nutritional value in its original stone-ground form, the benefits of whole grain were ignored for years by millers seeking to market a flour with uniform baking characteristics. Setting all nutrition considerations aside, bakers were able to present consumers with picture perfect loaves of white bread, but it cost the consumer the best part of the wheat kernel.

Fortunately, consumers now understand more about nutrition and diet than at any other time in history. Equally as fortunate, Howard and Anna Ruth Beck are helping us rediscover healthier ways to prepare meals. All of the recipes in *Whole Wheat Cookery* call for whole wheat flour complete with all the bran and wheat germ intact—nothing taken out, nothing added in.

Each of the 300-plus recipes in this cookbook uses stone-ground wheat flour or another wheat product. While the use of whole wheat flour conjures up thoughts of hearty breads, Anna Ruth has uncovered the unique nutty, crunchy texture and flavor for cakes, cookies and salads as well. We think you will find the freshness and variety in this cookbook quite delightful.

Stan Thiessen
Publisher

TABLE OF CONTENTS

PREFACE

As children growing up on wheat farms in Kansas, Howard and I knew the value of wheat grain to a healthy diet. Home made bread was an important part of every meal.

We continued our farming heritage after we were married and although farming in central Kansas is diverse, winter wheat was always the main crop.

We are retired now but farming has always had special significance to us because it was an inherited way of life. Howard and I were the third generation to till the same soil.

About seventeen years ago, Howard asked me to bake a loaf of bread from the wheat we had grown on our farm. He wanted to use it for a devotional service.

I began by grinding the wheat in a blender. It took several experiments but eventually, a good loaf of bread emerged from the oven. I have been experimenting with whole wheat flour ever since.

Whole wheat and other whole grains have a high nutritional value. As we shared our experiments and recipes with others, we discovered a market for ground whole wheat. Before long, we were milling, packaging and selling Stone Ground Whole Wheat Flour as well as Cracked Wheat Cereal and Whole Wheat All Purpose Baking Mix. Our company was simply named, THE WHEAT BIN and to-date, we have sold whole wheat products to Grocery Stores, Health Food Stores, Bakeries, Restaurants, Institutions, Gift and Specialty Shops, not to mention all the individual sales.

Through this cookbook, I hope to encourage you to try whole wheat recipes. You will probably begin some experimenting of your own. If your family is not familiar with the unique texture and flavor of foods made with stone ground whole wheat flour, cracked wheat, and/or cooked wheat kernels, you are in for a treat.

All of the recipes in this cookbook include wheat in some form and here is a small tip to get you going — home made wheat bread is the best for making dishes that call for bread crumbs.

Anna Ruth Beck

ACKNOWLEDGEMENTS

I would like to thank my husband, Howard, for his help, patience and encouragement while I worked on this cookbook and to our daughter, Marlene Yoder, for the many hours she spent typing recipes. Many family members shared recipes that appear in this cookbook. Other relatives and friends encouraged me to publish the *Whole Wheat Cookbook*. Some of their recipes are also included.

Magazines and newspapers provide a good source for recipes. I would like to thank all those who shared their favorite recipes in this way.

I owe a big THANK YOU to Ken Gingerich. He not only suggested I author the *Whole Wheat Cookbook* but offered his technical assistance in design and printing specifications.

STONE GROUND WHOLE WHEAT FLOUR

THE WHEAT BIN flour is stone ground and bagged without preservatives. Therefore, always store it in refrigerator or freezer in a plastic or metal container. The germ is broken down and it contains oil so will become rancid if not refrigerated. Few nutrients are lost in storing. Flour can be frozen for long periods of time.

Most of the vitamins and minerals are in the bran and germ. The endosperm contains almost exclusively carbohydrate (starch) and protein. THE WHEAT BIN's whole wheat flour contains all the vitamins and minerals since we grind the whole grain. "We add nothing to our products and we take nothing out."

When whole wheat flour is mentioned, the first thing we normally think of is whole wheat bread. However, quick breads, cakes, and cookies made with Stone Ground Whole Wheat Flour are also very delicious. Using whole wheat flour usually gives a heavier product. Some of us enjoy a heavier product, but many people are not used to it and therefore "do not like it." If your family is not used to using STONE GROUND WHOLE WHEAT FLOUR, begin by substituting a small amount for all-purpose/unbleached flour in pancakes, waffles, cookies, bread, etc. and gradually increase the amount of STONE GOUND WHOLE WHEAT FLOUR in recipes.

Some Baking Tips:

- Always stir STONE GROUND WHOLE WHEAT FLOUR before measuring.
- In yeast breads, add the STONE GROUND WHOLE WHEAT FLOUR to the liquid (if you use 100° whole wheat flour, add approximately half of it to the liquid), mix well, 3-4 min., and let stand for 15 minutes to let the gluten develop. Then add enough flour to finish the dough. Knead well, about 10 minutes. A lighter product will result if the gluten is developed.
- Bread flour may be used for a part, or all, of the white flour in yeast dough.
- Adding eggs and/or oatmeal to Whole Wheat bread makes it lighter and less crumbly.

- Use Lecithin (granules, liquid, or powder) to make a lighter bread.
- If the bread cracks on the sides, you have used too much flour. The dough will remain slightly sticky when hand kneading.
- Use oil on your hands when hand kneading the dough so you won't add too much flour.
- Place a pan of hot water in the bottom of the oven when you bake bread and also when you allow bread to rise in the oven to give the needed moisture to keep the loaves of bread from splitting on the sides and drying.

CRACKED WHEAT CEREAL

CRACKED WHEAT CEREAL is a very nutritious, high fiber cereal. It is easy to prepare as you will note in the following recipe:

1 c. cracked wheat | **2½ c. boiling water**

Add cracked wheat slowly to boiling water, stirring briskly. Cover and simmer 15 to 20 minutes (7 to 10 minutes if you prefer it more chewy). Let stand a few minutes. Serve with milk and honey OR brown sugar OR white sugar to taste.

CRACKED WHEAT is very compatible with ground beef and can be added to your favorite recipes.

CRACKED WHEAT may be substituted in recipes calling for bulgar.

There are a number of recipes in the Main Dishes and Side Dishes sections that use CRACKED WHEAT CEREAL. Also, some of the salads have it listed in the ingredients.

WHOLE WHEAT KERNELS

Cooked WHEAT KERNELS can also be used in a variety of ways. We like the nutty flavor they give to Salads and Desserts as well as in the Main Dishes and Side Dishes.

There are several ways to cook the WHEAT KERNELS so choose which is the most convenient for you from the following methods:

Note: Soaking WHEAT KERNELS overnight in the water it is to be cooked in, cuts cooking time in half in all the following methods.

STOVE TOP: 1 cup WHEAT KERNELS to 3 cups water. Simmer covered 1 hour, or 30 minutes if pre-soaked.

CROCK POT: 2 cups WHEAT KERNELS to 4 cups water. (Amounts can be doubled if you have a large crock pot.) Place WHEAT KERNELS and water in crock pot. Cover and cook on low approx. 6 hours. Stir once during the first hour of cooking. Refrigerate or freeze. Ready to use in your recipes.

OVEN-COOKED: 1 cup WHEAT KERNELS to 2 cups water. Preheat oven to 300° F. Boil in heavy saucepan 5 minutes. Remove from heat, cover and place in oven. Turn off oven heat. Leave undisturbed aprox. 6 hours.

One of the pluses of eating WHEAT is you don't have to eat so much to be satisfied, thus you are not eating so many calories!

BREADS

BLENDER POP-UP BREAD

1 c. hot tap water
½ c. powdered milk
¼ c. vegetable oil
¼ c. granulated sugar
1 tsp. salt
2 eggs
1 pkg. active dry yeast
1¼ c. all-purpose flour
1¾ c. stone ground whole wheat flour

Combine all ingredients except flour; blend 2 to 3 minutes. Stir blended ingredients into flour. Let rise 1½ hours. Grease two 1 lb. coffee cans and divide dough between the two cans. Cover with greased lids and allow to rise until the lids pop off the cans. Bake in preheated 375° oven for 35 minutes. Makes 2 loaves.

CRACKED WHEAT-HONEY BREAD

2 pkgs. dry yeast
2¼ c. warm water
½ c. honey
¼ c. shortening
2 tsp. salt
2 tblsp. Lecithin (granules, powder or liquid) optional
½ c. cracked wheat
4 c. stone ground whole wheat flour
2-3 c. unbleached flour

Dissolve yeast in ½ cup warm water in large mixing bowl. Add remaining ingredients, except unbleached flour. (Don't forget the 1¾ cups warm water.) Beat until smooth, about 7 minutes. Mix in unbleached flour to make dough easy to handle. Turn onto lightly floured board. Knead smooth, about 10 minutes. Place in lightly greased bowl. Turn greased side up. Cover; let rise until double in size, about 1 hour. Punch down; divide into halves. Form two loaves. Let rise until doubled in size, or until loaves sound hollow when tapped. Bake at 375° for 40-45 minutes. Cool on wire racks.

CRUSTY WHOLE-WHEAT OATMEAL BREAD

2 c. all-purpose flour
2 tsp. salt
2 pkgs. active dry yeast
3 c. water
⅓ c. molasses
6 tblsp. butter OR margarine
3 c. plus 2 tblsp. quick-cooking old-fashioned oats, uncooked
About 4½ c. stone ground whole wheat flour
1 egg

In large bowl, combine all-purpose flour, salt, and yeast. In 2-quart saucepan over low heat, heat water, molasses, butter or margarine, and 3 cups oats until very warm. (Butter or margarine does not need to melt completely.) With mixer at low speed, gradually beat oat mixture into dry ingredients just until blended. Increase speed to medium; beat 2 minutes, occasionally scraping bowl with rubber spatula. Beat in 1 cup whole wheat flour to make a thick batter; continue beating 2 minutes, occasionally scraping bowl. With wooden spoon, stir in 3 cups whole wheat flour to make a soft dough. Turn dough onto lightly floured suface. With floured hands, knead dough until smooth and elastic, about 10 minutes, working in about ½ cup whole wheat flour while kneading. Shape into ball and place in greased bowl turning dough over so top is greased. Cover with towel and let rise in warm place until doubled, about 1 hour. Grease large cookie sheet. Punch dough down. Shape into a French loaf and put on cookie sheet. Let rise until doubled, about 40 minutes. Preheat oven to 350°. With sharp knife, make slashes in top of loaf. In cup with fork, beat egg, brush bread with beaten egg; sprinkle with 2 tblsp. quick oats. Bake 1 hour or until bread sounds hollow when tapped with fingers. Remove from cookie sheet and cool on wire rack. Makes 1 large loaf or 2 small loaves.

DILLY BREAD

1 pkg. dry yeast
¼ c. water
1 c. creamed cottage cheese
1 tblsp. salad oil
½ tsp. soda
1 tsp. salt
2 tsp. sugar
1½ c. flour
1 c. stone ground whole wheat flour
1 egg, unbeaten
1 tsp. dill seed
1 tsp. onion flakes, caraway seed or a little garlic

Mix well and knead as usual. Let rise until light. Punch down well and place in one large or two small pans. Let rise again. Bake at 325° for 45 minutes for 1 loaf; 2 loaves - 350° for 30 minutes.

EARLY COLONIAL BREAD

1 c. yellow cornmeal
4 c. boiling water
½ c. honey
⅔ c. brown sugar
2 tblsp. salt
½ c. oil
4 pkgs. yeast
1 c. lukewarm water
½ c. oatmeal
½ c. wheat germ
1½ c. rye flour
2½ c. stone ground whole wheat flour
8 to 9 c. all-purpose flour

Combine cornmeal, whole wheat flour, sugar, salt, boiling water, oil and honey, mix well. Cool to lukewarm (about 30 minutes). Soften yeast in warm water. Then add to the cornmeal mixture. Now add the rest of the flours, oatmeal and wheat germ. Then add white flour and stir until the dough is stiff but not dry. Cover bowl and let rise in a warm oven until it has doubled in bulk (30-45 minutes). Take doubled dough, punch down and roll out on lightly floured surface. Cut into 4 equal parts. Knead each part 6-8 minutes. Shape into 4 loaves and place in greased 9 x 5 x 3-inch loaf pans. Cover and let rise again until doubled. Bake for 10 minutes in preheated 400° oven. Reduce heat to 325° and bake for 30 minutes longer. Remove from pans to cool.

HOLIDAY BREAD

2 pkg. active dry yeast
¼ c. lukewarm water
2 c. milk, scalded
⅔ c. shortening
⅔ c. sugar
2 tsp. salt
¼ tsp. ground cardamom
1 tsp. grated orange rind
2 beaten eggs
3-4 c. all-purpose flour
3 c. stone ground whole wheat flour

Soften yeast in warm water. Combine hot milk, shortening, sugar and salt. Cool to lukewarm. Add softened yeast. Add eggs, mix well. Gradually add flours, beating well. Let rise until double. Punch down. Divide dough in half. Let rest about 5 minutes. Roll out each part to make a long flat loaf 6 inch wide. Place on greased cookie sheets. Spread filling (recipe following) down center of each loaf. Cut dough in little strips on each side almost to filling and criss cross dough across filling. Let rise until doubled. Bake in 350° oven for 35 minutes.

Bread Filling

1 lb. pitted dates, chopped
½ c. brown sugar
½ c. water
½ c. chopped nuts
¼ c. maraschino cherries

Combine all ingredients and cook until thick. Cool before placing on dough.

HONEY-OATMEAL-WHEAT BREAD

3 c. boiling water
1½ c. old-fashioned oatmeal
½ c. lukewarm water
3 pkg. dry yeast
¾ c. honey
6 tblsp. cooking oil
3 tsp. salt
6 c. (about) stone ground whole wheat flour
3 c. (about) all-purpose flour

Place oatmeal in large mixing bowl. Pour boiling water over it; let cool until lukewarm. Dissolve yeast in ½ cup lukewarm water; let stand 5 minutes. Add dissolved yeast to oatmeal mixture. Add honey and oil; blend. Combine flours and salt. Add flours to oatmeal mixture. Mix thoroughly to form soft dough. Turn out onto lightly floured surface and knead 5 to 10 minutes. Form into loaves and place in greased pans. Let rise 30 minutes. Bake 1 hour in 325° oven. Cool on racks. For soft crust, brush loaf tops with butter while hot. Makes 2 or 3 loaves.

HONEY WHOLE-WHEAT BREAD

2 pkgs. dry yeast
½ c. warm water
⅓ c. honey
¼ c. shortening
1 tblsp. salt
1¾ c. warm water
3 c. stone ground whole wheat flour
3-4 c. all-purpose flour
1 tblsp. margarine, softened

Dissolve yeast in ½ cup warm water in large bowl. Stir in honey, shortening, salt, 1¾ c. warm water and whole wheat flour, beat until smooth. Fold in all-purpose flour to make dough easy to handle. (Start with 3 cups and add more as necessary.) Turn dough onto lightly floured surface and knead until smooth and elastic (about 10 minutes) or use dough hook. Place in greased bowl, turn greased side up. Let rise in warm place about an hour, until dough doubles. Punch dough down and form into 2 loaves. Place in 2 greased loaf pans. Brush with margarine and sprinkle with whole wheat flour or crushed oats. Let rise until double (about an hour). Heat oven to 375°. Bake until deep golden brown and sounds hollow when tapped, about 40 to 45 minutes. Remove from pans and cool on wire racks. Makes 2 loaves.

KANSAS WHEAT BREAD

1 pkg. dry yeast
2¼ c. warm water
1 tsp. sugar
2½ c. all-purpose flour
3 tblsp. margarine
2 tblsp. molasses
1½ tsp. salt
2 c. stone ground whole wheat flour
1-1½ c. additional whole wheat flour
1 c. raisins (optional)

Make a sponge by dissolving yeast in warm water. Add sugar and 2½ c. all-purpose flour. Mix well. Let rise until bubbly, about 20 minutes. Add margarine, molasses, salt and 2 c. whole wheat flour to sponge. Beat well. Knead in enough of the remaining whole wheat flour to make soft dough. Knead in raisins if used. Continue to knead until smooth and elastic. Let rise until doubled. Punch down. Divide in half. Roll and shape into 2 loaves. Place in greased 9 x 5 x 3-inch pans. Let rise. Bake in preheated 375° oven for 40 minutes or until crust sounds hollow when tapped. Remove immediately from pans and cool on wire rack. Makes 2 loaves.

PILGRIM'S BREAD

½ c. yellow cornmeal
⅓ c. packed brown sugar
1 tblsp. salt
2 c. boiling water
¼ c. cooking oil
2 pkg. dry yeast
¾ c. rye flour
2¼ c. stone ground whole wheat flour
2 c. unbleached flour
½ c. warm water

In a large mixing bowl, combine the cornmeal, brown sugar, salt, oil, and boiling water. Cool to lukewarm. Dissolve yeast in ½ c. warm water and add to cornmeal mixture. Beat in the whole wheat and rye flours, mixing well. Add enough unbleached flour to make a moderately stiff dough. Knead quickly until texture becomes smooth and elastic, 10-15 min. Place in greased bowl, turning so greased surface is on top. Cover and let rise until double. Shape into loaves. Let rise until doubled. Bake at 375° for 35 to 40 min. until browned and loaf sounds hollow when tapped.

RAISIN APPLE WHEAT BREAD

⅔ c. milk
¼ c. cracked wheat
2 pkgs. active dry yeast
½ c. warm milk
1 egg
2 tblsp. vegetable oil
2 tblsp. honey
2 tsp. salt
1 c. chopped tart apples
1½ c. all-purpose flour
2½ c. stone ground whole wheat flour
1½ c. raisins

Heat milk to lukewarm in small saucepan. Stir in cracked wheat, set aside. In large mixer bowl dissolve yeast in warm water. Add egg, oil, honey, salt, apple and milk mixture. Beat just to blend. Add the whole wheat flour. Beat at low speed just to blend, then beat at high speed for 5 minutes. With wooden spoon mix in ½ cup of the remaining flour and the raisins. Turn out onto floured board and knead 8 to 10 minutes, working in enough of the remaining flour to make a smooth, non-sticky dough. Place in greased bowl. Cover and let rise until double, about 1 - 1½ hours. Punch down, divide in half and form into 2 loaves. Place in 2 greased loaf pans. Cover and let rise in warm place until dough rises just above tops of pans, about 45 minutes to 1 hour. Bake in 350° oven 30-40 minutes until browned and loaves sound hollow when tapped. Cooled loaves may be wrapped and frozen. This bread makes excellent toast. Makes 2 loaves.

SWEDISH LIMPA

2½ c. very warm water
2 pkg. active dry yeast
¼ c. brown sugar
⅓ c. molasses
3 tblsp. soft shortening
1 tblsp. salt
2 tblsp. grated orange rind
1 tsp. anise seeds, crushed
1 c. cracked wheat
3½ c. rye flour
1½-2 c. all-purpose flour
2 c. stone ground whole wheat flour

In large mixing bowl dissolve yeast in ½ cup of the warm water. Add all the rest of the ingredients to the dissolved yeast except the all-purpose flour. Beat with electric mixer at medium speed for 10 minutes. Gradually stir in enough all-purpose flour to make a soft dough. Turn out onto a lightly floured surface. Knead until smooth and elastic. Be careful not to add too much flour. The dough should be rather sticky. Place dough in greased bowl; turn dough in bowl to grease top; cover with plastic wrap and a towel. Let rise until double or about 1½ hours. Punch dough down; turn out onto lightly floured surface; invert bowl over dough; allow to rest 10 minutes. Divide in half and knead each half a few times. Then shape each into an oval loaf. Place on greased cookie sheets OR you can shape into a loaf and put in loaf pans to bake. Let rise in a warm place 40 minutes or until double in size (brush with egg to give a shiny top - 1 egg yolk plus 2 tblsp. water). Bake at 375° for 45 minutes or until golden brown and loaves sound hollow when tapped. Remove from pans and cool on wire rack.

WHOLE WHEAT DOUGH BASIC RECIPE

1½ c. warm water
6 tblsp. dry yeast
Pinch sugar
2 c. warm water
1 tblsp. salt
1 c. salad oil
¼ c. honey
3 eggs, beaten
10½ c. stone ground whole wheat flour

Add yeast and pinch of sugar to 1½ c. warm water and let set till bubbly. Combine in mixer bowl which uses bread dough hook; remaining water, salt, oil, honey, eggs, flour and yeast mixture. Turn mixer on low and knead for 10 minutes. Remove dough hook from dough and let the dough rise in the mixer bowl till double. Punch down and use in the following ways:

Pizza Crust: Spread 1-1½ cups dough on greased cookie sheet, using rolling pin to spread. Fill with favorite pizza sauce and topping. Bake in 400° oven for 20 minutes.

Hot Dog & Hamburger Buns: If you do not have specially made pans to make the desired buns, you may use greased jelly roll pans. Use enough dough to make the size of bun wanted, about ½ - ¾ cup. Shape and place 1-2 inches apart in pan. In blender, mix 1 egg white and 1 tsp. water. Brush on top of buns and sprinkle with hulled sesame seeds. Let rise slightly and bake in 400° oven for 10 minutes or until desired browness.

Clover Leaf Rolls: Grease muffin cups. Shape dough into 1-inch balls and place 3 balls into each cup. Let rise till double. Bake at 400° for 15 minutes. Let stand for 5 minutes.

Cinnamon Rolls: Roll half of the basic recipe out on floured board until 8 inches wide and ¼ inch thick. Combine ½ cup sugar, ¼ c. melted or soft butter, and 1½-2 tsp. cinnamon and spread over dough. Sprinkle with ¼-½ cup raisins. Roll up like jelly roll. Cut the dough in 1-1½-inch pieces. Let rise till almost double. Bake at 375° until golden brown.

100% WHOLE WHEAT BREAD

3 c. scalded milk, cooled to lukewarm or 3 c. lukewarm water (plus ¾ c. dry milk powder, optional)
2 tblsp. yeast
¼ c. honey
2½ tsp. salt
¼ c. oil or melted butter
7-8 c. stone ground whole wheat flour
1-2 c. stone ground whole wheat flour for kneading

Combine milk or water and honey; stir well. Sprinkle yeast over sweetened liquid and let set until bubbly (about 10 minutes). Next, add 3-4 cups of the flour. Beat it in well and vigorously. Let this mixture rise in a warm spot for about 20 minutes. Then add oil and salt, beating thoroughly, and add the additional flour by the cupful, beating vigorously until the dough is too stiff to beat. Turn dough out onto a large, clean, floured surface and knead the dough, incorporating flour as necessary; then place kneaded dough in a large, buttered bowl. Cover with a clean, damp towel and set the dough to rise in a warm spot until doubled, about 50-70 minutes. If you have time, punch the dough with your fist about 20 times after this first rising, then cover and allow to rise again. If your're short for time, just divide the dough into two and gently knead the dough into two loaf shapes. Any seam or uneven spot will be the bottom of the loaf. Decide which side is the bottom and poke that side in about 10 places with the tines of a fork. This will prevent any large air pockets from forming in the bread. Place in two well-buttered bread pans and let rise 20 minutes, covered. Preheat oven to 375°. Just before baking, make two or three diagonal slits in the top of the bread with a knife. This will allow the steam to escape. Brush top of loaves with a beaten egg mixed with a bit of milk if a shiny, golden crust is desired. Bake 45-50 minutes at 375°. Let bread cool thoroughly on rack before slicing. Store in a plastic bag when cool.

WHOLE WHEAT BREAD

3 tsp. sugar
1 c. warm water
4 pkgs. dry yeast
¾ c. sugar
½ c. oatmeal
½ c. cornmeal
1½ tblsp. salt
5¼ tblsp. melted butter OR oil
4½ c. warm water
9 c. stone ground whole wheat flour
6-7 c. all-purpose flour

Mix 3 tsp. sugar, 1 c. water, yeast together and let rise until double, 15-20 minutes. Combine ¾ c. sugar, 1½ tblsp. salt, oatmeal, cornmeal, melted butter and 4½ c. water. Add yeast mixture. Measure whole wheat flour into mixture. Beat with electric mixer 3 minutes. Add all-purpose flour. Dough should be smooth and just stiff enough not to stick to your hands after kneading 5 minutes. Place in bowl and warm place and let rise 30 minutes. Punch down and let rise another 30 minutes. Punch down and let rise 30 minutes again. Shape into loaves and punch with fork to get out air bubbles. Let rise 1 hour. Bake at 350° for 40-50 minutes. Remove from pan immediately and brush with margarine for soft crust bread.

WHOLE WHEAT BREAD

2 pkgs. active dry yeast
½ c. warm water
½ c. brown sugar, packed
1 tblsp. salt
¼ c. shortening
2¼ c. warm water
6-7 c. stone ground whole wheat flour

Dissolve yeast in ½ cup. warm water. Stir in brown sugar, salt, shortening, 2¼ cups warm water and 3½ cups of the flour. Beat until smooth. Mix in enough remaining flour to make dough easy to handle. Turn onto lightly floured board, knead until smooth and elastic, about 10 minutes. Place in greased bowl, turn greased side up. Cover; let rise in warm place until double, about 1 hour. Punch down, divide in half. Roll each half into rectangle 9 x 18-inches. Roll up beginning at short side. With side of hand, press each end to seal. Place seam side down in greased baking pan approximately 9 x 5 x 3-inch. Cover; let rise until double, about 1 hour. Heat oven to 375°. Place loaves on low rack so that tops of pans are in center of oven. Pans should not touch each other or sides of oven. Bake 40-45 minutes or until deep golden brown and loaves sound hollow when tapped. Remove from pans. Cool on wire rack. Makes 2 loaves.

WHOLE WHEAT EGG BREAD

2 c. boiling water
½ c. powdered milk
½ c. butter OR margarine
2 tsp. salt
½ c. honey
2 pkgs. yeast
2 eggs, beaten
4 c. stone ground whole wheat flour
4½ c. all-purpose flour

Pour boiling water over milk, butter, salt and honey. Allow to cool. Add yeast and eggs. Add flour, 1 cup at a time, starting with the whole wheat first. Turn dough onto floured board and knead until smooth and elastic. Place dough in a greased bowl, cover, and let rise until double in bulk (about 1½ hours). Punch down and turn onto a lightly floured board. Shape into five round loaves and place in greased pie pans. Cover and let rise until double again. Slit top (make an X) with a very sharp (serrated) knife. Bake in 350° oven for 25-30 minutes.
Variations: Add 1 cup of raisins for raisin bread; or make rolls.

WHOLE WHEAT POTATO-ONION BREAD

1¾ c. all-purpose flour
2 pkgs. active dry yeast
1 c. milk
½ c. water
2 tblsp. butter
2 tblsp. sugar
1 tblsp. salt
1½ c. seasoned mashed potatoes
½ c. dairy sour cream
½ c. minced onion
2 tsp. tarragon
1 tsp. garlic powder
5-6 c. stone ground whole wheat flour

Stir together 1¾ c. flour and yeast. Heat milk, water, butter, sugar and salt over low heat until warm, stirring to blend. Add liquid ingredients to flour-yeast mixture and beat until smooth, about 3 minutes. Add potatoes, sour cream, onion, tarragon and garlic powder. Beat until smooth. Stir in enough of the whole wheat flour to make a moderately stiff dough. Knead 10 minutes or until smooth. Turn out on floured surface; let rise in warm place until doubled, about 45-50 minutes. Punch down. Divide in half; shape into two loaves. Place in two greased loaf pans. Let dough rise again until almost doubled in volume, about 30 minutes. Bake in preheated 375° oven 35-40 minutes. Makes 2 loaves.

HARD ROLLS

- 2½ c. stone ground whole wheat flour
- 2-3 c. all-purpose flour
- 2 tblsp. sugar
- 2 tsp. salt
- 1 pkg. active dry yeast
- 3 tblsp. softened margarine
- 1½ c. very hot water
- 1 egg white (room temperature)
- Cornmeal
- ½ c. water
- 1 tsp. cornstarch

In large bowl thoroughly mix 1⅓ c. flour, sugar, salt and undissolved yeast. Add margarine. Gradually add hot water to dry ingredients and beat 2 minutes at medium speed, scraping bowl occasionally. Add egg white and 1 cup flour to make a thick batter. Beat at high speed 2 minutes, scraping bowl. Stir in enough additional flour to make a soft dough. Turn onto lightly floured board and knead until smooth and elastic. Place in a greased bowl, turning to grease top. Cover with plastic wrap and let rise in warm place until doubled. Punch down and turn out onto lightly floured board. Cover and let rest 10 minutes. Divide in half. Cut each half into nine 1-inch pieces. Form each into smooth balls. Place 3 inches apart on greased baking sheets that have been sprinkled lightly with cornmeal. Cover, let rise until doubled. Slowly blend ½ cup water into cornstarch in small pan and bring to boil until clear. Cool slightly. When ready to bake rolls, brush each with cornstarch glaze. Slit tops with a sharp knife-criss-cross fashion. Leave plain or sprinkle with poppy or sesame seeds. Bake at 450° about 15 minutes. Remove from baking sheets and cool on wire racks. Variation: Dough can be divided in half for two medium-sized loaves of bread. Bake about 30 minutes or until crusts sound hollow when tapped.

HONEY-WHEAT BREAD STICKS

- 1 pkg. dry yeast
- 1 c. warm water
- 1 c. shortening OR oil
- 3 tblsp., plus 1 tsp. honey, divided
- 2 tsp. salt
- 1 c. water, boiling
- 2 eggs, beaten
- 6 c. stone ground whole wheat flour

Dissolve yeast in warm water; set aside. In large bowl combine shortening or oil, 3 tblsp. honey, salt and boiling water; allow to cool to lukewarm. Add beaten eggs, 1 tsp. honey and yeast mixture. Add whole wheat flour. Mix well, but do not knead. Chill several hours or overnight. Divide into about 60 balls; form into pencil-sized sticks about 8 inches long. Cover; let rise on greased baking sheets until doubled. Bake at 325° for 30 minutes or until golden.

KANSAS WHEAT DINNER ROLLS

2 c. water
¾ c. shortening OR oil
2 tblsp. molasses
2¾-3¼ c. all-purpose flour
4 c. stone ground whole wheat flour
½ c. sugar
2 tsp. salt
2 pkg. active dry yeast
2 eggs

In small saucepan, heat first 3 ingredients until very warm. In large bowl, blend warm liquid, whole wheat flour, sugar, salt, yeast and eggs at low speed until moistened. Beat 8 minutes at medium speed. Stir in enough all-purpose flour to form a stiff dough. Knead on floured surface until smooth and elastic, about 5 minutes. Place dough in greased bowl, turn greased side up. Cover; let rise in warm place until doubled in size, 45-60 minutes. Generously grease one 9 x 13-inch pan and one 8-inch square pan. Punch down dough. Divide dough into 36 pieces, shape into balls. Place in prepared pans. Cover; let rise in warm place until doubled in size, 30-45 minutes. Heat oven to 375°. Bake 15-20 minutes or until golden brown and rolls sound hollow when lightly tapped. Immediately remove from pan; cool.

NO-KNEAD WHOLE-WHEAT ROLLS

1 pkg. yeast
¼ c. lukewarm water
1 c. milk, scalded*
½ c. butter OR margarine
¼ c. sugar
1 tsp. salt
1¾ c. stone ground whole wheat flour
1¾ c. all-purpose flour
1 well beaten egg

Dissolve yeast in water. Pour hot milk over margarine, sugar and salt. Let cool. Add whole wheat flour. Beat about 7 minutes. Add yeast and egg. Beat with mixer until smooth. Add approximately 1¾ cups all-purpose flour. (Add more flour if dough is too soft, although dough should be somewhat soft). Let rise until bubbly. Stir down, and let rise again. Stir down, and shape into rolls. Use for sweet or dinner rolls. *Sometimes I use dry milk in place of whole milk. Add dry milk to sugar and be sure to add another cup water.

WHOLE WHEAT CRESCENTS

1 pkg. yeast
1 c. warm water
3/4 c. evaporated milk
1 1/2 tsp. salt
1/3 c. honey
1 egg, beaten
2 c. whole wheat kernels
1/4 c. melted margarine
1 c. stone ground whole wheat flour
3 c. all-purpose flour
1 c. margarine

Dissolve yeast in warm water. Add evaporated milk, salt, honey, egg and whole wheat. Beat until smooth. Stir in melted margarine and set aside. In another bowl, stir together whole wheat flour and all-purpose flour. Cut in 1 c. margarine. Pour yeast batter into margarine-flour mixture and gently stir just until all flour is moistened. Divide into 4 parts and wrap in plastic wrap and refrigerate overnight. Roll out and form in crescent rolls and let rise—about 2 hours. Do not place in warm spot. Bake 325° for 25 minutes.

30-MINUTE HAMBURGER BUNS

3 1/2 c. warm water
1 c. oil
1/2 c. honey
6 tblsp. yeast
3 eggs
1 tblsp. salt
10 1/2 c. stone ground whole wheat flour
1/2 c. bran

Mix first 4 ingredients together and let rest for 5 minutes. Add the rest of the ingredients and knead until, when touched lightly, it springs back, about 10 minutes. Shape into rolls by rolling a ball of dough about the size of an egg and flatten. Place on greased cookie sheet. Let rise about 10 minutes and bake 10 minutes at 425°. Makes 4 dozen rolls.

ALL-SEASON QUICK LOAVES

1½ c. stone ground whole wheat flour
1¼ c. all-purpose flour
1 tsp. salt
1 tsp. soda
½ tsp. baking powder
2 tsp. cinnamon
3 eggs
1½ c. sugar
1 c. salad oil
1 tsp. vanilla
1 c. chopped nuts
2 c. prepared fruit or vegetable

Stir together the flours, salt, soda, baking powder, cinnamon and nuts; set aside. In a medium-size bowl, lightly beat the eggs. Add the sugar and oil and stir until blended. Stir in the vanilla and choice of fruit or vegetable. Add the flour mixture all at once and stir just until evenly moist. Then divide batter evenly between 2 greased and floured 4½ x 8½-inch loaf pans. Bake bread in a 350° oven for 50-60 minutes, or until a wooden pick inserted in the center comes out clean. Let stand 10 minutes, then turn out on a wire rack to cool completely. These breads are easier to slice if they sit at least a day. Wrapped well, they keep, refrigerated, up to a week. You can freeze them for longer storage. Makes 2 loaves. Fruit and Vegetable variations:

Orange Bread—Grate 1 tblsp. orange peel. Remove peel and remaining white membrane from 4 oranges. Remove seeds then chop oranges finely to make 2 cups. Combine with grated peel and add to bread mixture.

Apple Bread—Peel, core and shred 3 or 4 medium tart apples to make 2 cups total. Stir 1 tsp. lemon juice into apples.

Tomato Bread—Peel 3 or 4 medium tomatoes. Cut each in half and squeeze gently to remove seed pockets. Finely chop the firm pulp to make 2 cups total.

Zucchini Bread—Coarsely shred about 2 medium zucchini to make 2 cups total, packing lightly in measuring cup.

Patience and delay achieve more than force and rage.

APPLE CHEESE BREAD

- 2/3 c. sugar
- 1/2 c. butter OR margarine
- 2 eggs
- 1 c. stone ground whole wheat flour
- 1 c. all-purpose flour
- 1/2 c. chopped pecans or walnuts
- 1/2 c. shredded cheddar cheese
- 1 1/2 c. peeled and grated apples
- 1 tsp. baking powder
- 1 tsp. cinnamon
- 1/2 tsp. salt
- 1/2 tsp. baking soda
- 1/4 tsp. ginger

Heat oven to 350°. Beat together at medium speed sugar, butter and eggs, scraping down sides often until light and fluffy (2-3 minutes). Sift together dry ingredients and add to first mixture along with the cheese, nuts and apples. Blend well (the batter will be stiff). Spread into a greased 9 x 13-inch pan. Bake for 50-60 minutes or until wooden pick inserted in center comes out clean. Cool 10 minutes; remove from pan, cool completely.

APPLESAUCE-FRUIT BREAD

- 1/2 c. margarine
- 1 c. sugar
- 2 eggs
- 1 tsp. salt
- 1 1/2 c. sweetened applesauce
- 2 tsp. baking soda
- 1 c. nuts, chopped
- 1 tsp. cinnamon
- 1/2 tsp. cloves
- 1/2 tsp. nutmeg
- 1 c. all-purpose flour
- 1 c. stone ground whole wheat flour
- 1 c. raisins
- 1 c. chopped candied fruit

Pour boiling water over raisins and drain well before using. Cream together margarine and sugar. Add eggs and beat well. Mix applesauce and soda and add to sugar mixture. Add salt and chopped fruit. Sift spices with flour and add alternately with raisins. Beat just until flour is blended in. Bake in two greased and floured loaf pans. Bake at 350° 1 hour or until toothpick comes out dry when inserted in center of breads. Cool slightly, turn out onto a cooling rack. May be decorated with a powdered sugar icing. Additional nuts and candied fruit may be added for a fruit cake. Also may be baked in a greased and floured angel food pan.

APRICOT NUT BREAD

1 can (1 lb.) apricots
1/3 c. shortening
1/2 c. sugar
2 eggs
3/4 c. stone ground whole wheat flour
3/4 c. all-purpose flour
1 tsp. baking soda
1/2 tsp. salt
1/2 c. chopped walnuts

Preheat oven to 350°. Drain apricots, reserving syrup; sieve the apricots to make 1 cup of juice and pulp. Cream shortening and sugar until light and fluffy. Add eggs. Lightly stir dry ingredients together and add to creamed mixture alternately with apricot juice and pulp. Stir until just combined, then stir in nuts. Dried apricots can be added for extra flavor. Bake in well greased loaf pan for 50 minutes. If small loaf pans are used cut the baking time to 40 minutes.

BANANA MOLASSES BREAD

3 ripe bananas
1 egg, unbeaten
2/3 c. sugar
2 tblsp. molasses
2 tblsp. melted shortening
1 c. stone ground whole wheat flour
1 c. all-purpose flour
1 1/2 tsp. baking powder
1 1/2 tsp. baking soda
1/2 tsp. salt
1 c. chopped walnuts

Mash bananas until no lumps remain. Add egg and mix well, Beat in sugar, molasses, and shortening and then stir in walnuts. Mix and sift flour, baking powder, soda, and salt. Blend in. Turn into greased loaf pan 8 x 5 x 3-inches and bake in moderate oven at 325° for about 1 hour.

BANANA NUT BREAD

1 c. sugar
1/2 c. shortening
4 tblsp. sour milk
2 eggs, beaten
1 c. stone ground whole wheat flour
1 c. all-purpose flour
1/2 c. nuts
1 tsp. baking powder
1 tsp. soda
1 tsp. vanilla
3 mashed bananas

Stir all ingredients together. Put in a loaf pan and bake for approximately 1 hour at 350°.

BRAN BANANA NUT BREAD

¾ c. all-purpose flour
¾ c. stone ground whole wheat flour
2 tsp. baking powder
½ tsp. baking soda
½ tsp. salt
¼ c. soft shortening
½ c. sugar
1 egg
1½ c. mashed ripe bananas
1 c. whole bran cereal
1 tsp. vanilla flavoring
½ c. coarsely chopped nuts

Sift together flour, baking powder, soda and salt. Set aside. Measure shortening and sugar into mixing bowl, beat until light and fluffy. Add egg and beat well. Stir in bananas, cereal, vanilla and nuts. Add sifted dry ingredients; stir only until combined. Spread in well-greased 9 x 5 x 3-inch loaf pan. Bake in moderate oven (350°) about 1 hour or until done. Let stand in pan on cooling rack about 5 minutes. Invert pan to remove loaf. Cool thoroughly before slicing.

BUGS BUNNY BREAD

1½ c. finely grated carrots
½ c. nutmeats
½ c. raisins
1 tsp. cinnamon
¾ tsp. nutmeg
¼ tsp. salt
1 tsp. baking soda
1 ⅔ c. stone ground whole wheat flour
1 c. sugar or packed brown sugar
2 eggs
⅔ c. salad oil

Cream sugar with oil, then add eggs and cream well. Sift all dry ingredients, and add to creamed ingredients, blending well. Add carrots, mixing well. Stir in raisins and nutmeats. Grease standard loaf pan, and pour batter in. Bake in 350° oven for 1 hour.

CARROT-BANANA BREAD

¼ c. margarine
½ c. brown sugar
2 eggs
1 c. mashed banana
1 c. stone ground whole wheat flour
1 c. all-purpose flour
1 tsp. baking soda
1 tsp. baking powder
½ tsp. cinnamon
¼ tsp. salt
1 c. grated carrots
½ c. chopped nuts, if desired

Cream margarine and sugar. Add eggs and mashed bananas. Sift dry ingredients together and add to creamed mixture. Stir in carrots and nuts. Bake in oiled 9 x 5 x 3-inch pan at 350° for 50-60 minutes. Cool 10 minutes. Turn out of pan to cool completely.

CARROT-NUT BREAD

1 c. sugar
1/3 c. vegetable oil
2 eggs
3/4 c. all-purpose flour
3/4 c. stone ground whole wheat flour
1/2 tsp. salt
1 tsp baking soda
1 tsp. baking powder
1 tsp. cinnamon
1 c. grated carrot
1/2 c. chopped walnuts or pecans

Preheat oven to 375°. Grease an 8 x 4-inch loaf pan. In medium bowl, combine sugar, oil and eggs. Stir to blend. Blend flours, salt, baking soda, baking powder and cinnamon. Stir into egg mixture until dry ingredients are moistened. Stir in carrots and nuts. Spoon mixture into greased pan; smooth top. Bake in preheated oven 50 to 55 minutes or until a wooden pick inserted in center comes out clean. Cool in pan on wire rack 5 minutes; remove from pan, cool on rack. Makes 1 loaf.

CARROT PINEAPPLE RING

1 1/4 c. stone ground whole wheat flour
1 tsp. baking soda
1/4 tsp. salt
1/4 tsp. cinnamon
1/4 tsp. nutmeg
1/4 tsp. all-spice
1/4 tsp. cloves
1/2 lb. carrots
2 eggs, at room temperature
3/4 c. unsalted sweet butter, melted, hot
1 1/4 c. sugar
1/3 c. packed flaked coconut
1 tsp. vanilla
1 c. walnuts
1/2 c. drained, canned crushed pineapple

Combine flour, salt, cinnamon, nutmeg, all-spice and cloves in dry food processor container fitted with metal blade. Process with 2 five-second on/off turns. Transfer to waxed paper. Cut carrots into even lengths to fit upright in food chute. Shred carrots with shredding disc, transfer to waxed paper. Process eggs in food processor container fitted with metal blade, adding butter in thin stream within 20 seconds; process 15 seconds. Process, gradually adding sugar withing 15 seconds. Scrape down side of container. Add carrots, coconut and vanilla; process with 4 half-second on/off turns. Add flour mixture, walnuts and pineapple; process with 6 or 7 half-second on/off turns just until flour is mixed into batter. Pour batter into buttered 8 cup ring mold. Bake at 350° until wooden pick inserted in center is withdrawn clean, about 40 minutes. Cool. Remove cake from pan.

CRANBERRY BREAD

- 2 c. stone ground whole wheat flour
- 2 c. all-purpose flour
- 1 tblsp. baking powder
- 1 tsp. baking soda
- 1 tsp. salt
- ½ tsp. cinnamon
- ¼ tsp. nutmeg
- 1 c. brown sugar
- 1 c. sugar
- ½ c. butter
- 1 tblsp. grated orange rind
- 1½ c. orange juice
- 2 eggs
- 2 c. fresh cranberries, coarsely chopped
- 1 c. chopped nuts
- ⅔ c. raisins

Sift dry ingredients together. Cut in butter until mixture resembles coarse meal. Combine orange rind, orange juice and eggs; add to dry ingredients, mixing just to moisten. Fold in berries, nuts, raisins. Turn into 2 greased and floured 9 x 5-inch loaf pans. Bake in 350° oven for 55-60 minutes. Cool. Makes 2 loaves. Slices better the next day.

HONEY-SPICED WHEAT BREAD

- ⅓ c. honey
- 1⅓ c. water
- 3 tblsp. oil
- 2 tsp. grated orange rind
- ¾ tsp. salt
- 1 tsp. cinnamon
- Pinch powdered cloves
- Pinch anise seed
- ⅛ tsp. ginger
- ⅛ tsp. allspice
- 2 tblsp. lemon juice
- 1 c. stone ground whole wheat flour
- 1½ c. whole rye flour
- 2½ tsp. baking powder
- ½ tsp. baking soda
- ½ c. chopped toasted almonds

Preheat oven to 350°. Grease an 8 x 4-inch loaf pan or an 8 x 8-inch pan. Combine the honey, water, oil, orange rind, salt, cinnamon, cloves, anise, ginger, allspice, and lemon juice. Sift together the flours, baking powder and soda. Add them gradually to the liquid ingredients and stir in the almonds, reserving 2 tablespoons to sprinkle on top. Bake 45 minutes to 1 hour, or 30 minutes if the pan is flatter. Cool at least 10 minutes before slicing or cutting in squares. Makes one loaf.

JALAPENO WHEAT/CORN BREAD RING

Shortening
¼ c. cornflakes crumbs
½ tsp. chili powder
½ c. margarine
2 tblsp. sugar
2 eggs
1 c. milk
1 c. stone ground whole wheat flour
1 c. yellow cornmeal
1 tblsp. baking powder
½ tsp. salt
2 jalapeno peppers, seeded and minced
1 can (16 oz.) chopped green chilies, drained
½ c. shredded sharp cheddar cheese

Using solid shortening, grease an 8 or 12 cup ring pan. Combine cornflakes crumbs and chili powder, pour into pan and tilt to coat. Place margarine in a glass mixing bowl and microwave on high 1 minute. Beat in sugar and eggs; then blend in milk. Stir in flour, cornmeal, baking powder and salt. Fold in peppers, green chilies and cheese. Pour into prepared pan; level batter. Microwave on 50 percent (medium) 5 minutes. Rotate pan; then microwave on high 3 to 5 minutes, or until corn bread tests done. Let stand 3 minutes before inverting onto service platter. Makes 16 slices.

JIFFY FRUIT BREAD

½ c. butter OR margarine
1 c. sugar
2 eggs
1 c. mashed ripe bananas
1 c. all-purpose flour
1 c. stone ground whole wheat flour
1½ tsp. soda
½ c. chopped nuts
½ c. chocolate chips
½ c. maraschino cherries, sliced

Cream butter and sugar. Add eggs. Continue beating until fluffy. Combine flours and soda. Add to other ingredients. Stir in nuts, chips and cherries. Pour into buttered and floured 5 x 9-inch loaf pan. Bake at 350° for 50-60 minutes. Do not double the recipe—repeat it instead.

The peace of God passeth all understanding and misunderstanding.

PUMPKIN BREAD

1¾ c. stone ground whole wheat flour
1⅓ c. all-purpose flour, sifted
½ tsp. baking powder
2 tsp. soda
1½ tsp. salt
1½ tsp. cinnamon
¼ tsp. cloves
⅔ c. chopped nuts
⅔ c. raisins or dates
2 c. sugar
⅔ c. salad oil or shortening
4 eggs
2 c. canned pumpkin
⅔ c. water

Preheat oven to 350°. Stir together the flour, baking powder, soda, salt, and spices. Toss nuts and fruit lightly with flour mixture. Beat sugar and oil in mixer at slow speed. Add eggs one at a time. Add pumpkin. Stir in flour mixture and water. Mix thoroughly. Pour into two greased loaf pans. Bake for 1 hour. Makes 3 medium loaves or two large loaves.

SPICY APPLE BREAD

1 c. stone ground whole wheat flour
1 c. all-purpose flour
1 tsp. baking soda
½ tsp. nutmeg
½ tsp. cloves
½ c. shortening
¾ c. firmly packed brown sugar
2 eggs
1 c. coarsely grated raw apple
¼ c. buttermilk
½ c. chopped nuts

Preheat oven to 350°. Sift the dry ingredients together and set aside. Combine shortening, brown sugar, and eggs, and beat well. Stir in half of the dry ingredients and apples. Add buttermilk and blend thoroughly. Add remaining dry mixture and stir until just well mixed. Stir in nuts. Bake in a greased loaf pan for 55 to 60 minutes. Turn out on rack and cool thoroughly.

RHUBARB BREAD

1½ c. granulated or brown sugar
⅔ c. vegetable oil
1 egg
1½ c. stone ground whole wheat flour
1¼ c. all-purpose flour
1½ tsp. baking soda
1 tsp. salt
1 c. sour milk OR buttermilk
2 c. rhubarb, finely cut
½ c. chopped nuts, optional
Topping:
½ c. brown sugar
2 tblsp. stone ground whole wheat flour
½ tsp. cinnamon
2 tblsp. melted butter OR margarine

Preheat oven to 325°. Beat together sugar, oil and egg. Stir together flour, baking soda and salt; add alternately with milk to oil mixture. Do not overmix. Stir in rhubarb and nuts. Pour into greased bread pan. Topping: Combine dry ingredients and using a fork stir in melted butter. Sprinkle topping onto batter surface and press lightly. Bake 60 minutes or until it tests done.

SQUMPKIN BREAD

3¼ c. stone ground whole wheat flour
1½ c. sugar
1½ tsp. baking soda
1½ tsp. salt
1 tsp. ground cinnamon
1 tsp. ground nutmeg
1 c. cooking oil
4 eggs
2 c. mashed, cooked butternut squash
⅔ c. water

Sift together flour, sugar, baking soda, salt, cinnamon and nutmeg into bowl, set aside. Combine oil, eggs, squash and water in bowl; blend well. Add squash mixture all at once to dry ingredients, stirring just enough to moisten. Pour batter into 2 greased 9 x 5 x 3-inch loaf pans lined with waxed paper. Bake in 350° oven 1 hour 15 minutes or until a wooden pick comes out clean. Cool completely. Makes 2 loaves.

Life gives many second chances — if you have the ability to recognize them and the courage to act.

VEGETABLE HEALTH BREAD

½ c. grated carrots
½ c. grated apple
½ c. grated potato
½ c. grated zucchini
1½ c. stone ground whole wheat flour
¾ c. sugar
1 tsp. cinnamon
¼ tsp. nutmeg
1½ tsp. baking powder
½ tsp. baking soda
½ tsp. salt
½ c. oil
1 egg
2 tblsp. cider OR water
½ c. nutmeats
1 c. raisins

Combine dry ingredients. Add oil, egg and cider or water and mix well. Add vegetables, nuts and raisins. Bake in greased 2 small or 1 large loaf pan in 350° oven. Freezes well and is better after the first day.

WHEAT FOLD-OVER CRESCENTS

½ c. cooked wheat kernels
2 tblsp. favorite jam
1½ c. stone ground whole wheat flour
1 tsp. baking powder
¼ tsp. salt
1 tsp. baking soda
½ c. butter OR margarine
½ c. brown sugar
1 egg
1 tsp. vanilla

Filling: Combine wheat and jam. Cover and refrigerate. Dough: Stir together the dry ingredients in small bowl. Cream margarine or butter. Beat in sugar until fluffy. Add the egg and vanilla and beat well. Add dry ingredients and blend well. Cover and chill for 2 hours. Roll out dough to ⅛-inch thickness and cut in 2½-inch circles. Place 1 teaspoon filling in center of each circle and fold in half shaping into crescents. Seal edges with a fork. Bake on ungreased baking sheet at 375° for 10 to 12 minutes. Cool on rack. Makes 24.

The soul is dyed with the color of its leisure thoughts.

WHOLE WHEAT ZUCCHINI BREAD

3 eggs, beaten
1 c. sugar
1 tsp. vanilla
1/3 c. honey
1 c. salad oil
2 1/2 c. stone ground whole wheat flour
2 tsp. baking soda
1 tsp. baking powder
1/2 tsp. salt
1 c. chopped nuts
2 c. grated zucchini
3 tsp. cinnamon

Beat eggs, sugar, vanilla, honey and salad oil together. Sift dry ingredients together and add to first mixture. Mix well, then fold in nuts and zucchini. Pour into 3 greased and floured loaf pans. Bake at 350° about 40 minutes or until cake tester comes out clean. Cool in pans for 10 minutes, then turn out to complete cooling.

ZUCCHINI BREAD WITH PINEAPPLE

3 eggs, beaten
1 c. oil
1 c. sugar
3 tsp. vanilla
1 c. raisins, chopped
1 c. nuts
2 c. grated zucchini
1 c. crushed pineapple, drained
1 1/2 c. stone ground whole wheat flour
1 1/2 c. all-purpose flour
2 tsp. soda
1 1/2 tsp. cinnamon
1 tsp. salt

Mix flour, soda and raisins and add to other ingredients. Bake at 350° for 1 hour. Makes 4 medium loaves or 1 9 x 13-inch cake pan.

WHOLE-WHEAT BROWN BREAD

1 c. all-purpose flour
1 1/2 c. stone ground whole wheat flour
1 1/2 tsp. baking powder
1 tsp. soda
1 tsp. salt
1 1/2 c. buttermilk
1/4 c. cooking oil
1/2 c. honey (OR 1/4 c. honey and 1/4 c. molasses)

Combine all ingredients. Pour into greased 4 1/2 x 8-inch loaf pan. Bake 40-50 minutes at 350°. (No eggs are needed in this batter.)

WHOLE-WHEAT FRUIT BREAD

1 egg
¼ c. honey
¼ c. maple syrup
1⅓ c. milk
3 tblsp. salad oil
1 c. chopped dates
½ c. chopped walnuts
1 c. flaked coconut
½ c. raisins
2 tsp. grated orange rind
1½ c. all-purpose flour
¼ tsp. baking soda
3 tsp. baking powder
1 tsp. salt
1 c. stone ground whole wheat flour
½ c. wheat germ

Beat together the first 3 ingredients. Add milk and salad oil. Add dates, nuts, coconut, raisins and orange rind. Sift together flour, soda, baking powder and salt. Add whole wheat flour and wheat germ. Stir into first mixture carefully, mixing just enough to dampen flour. Turn into a greased and floured 9 x 5 x 3-inch loaf pan or 2 4 x 7½ x 2½-inch pans. Bake at 325° for 1 hour and 15 minutes for the large loaf or 1 hour for the small loaves.

WHOLE WHEAT NUT BREAD

¾ c. all-purpose flour
3 tsp. baking powder
¼ tsp. salt
½ tsp. cinnamon
¼ tsp. nutmeg
¼ tsp. allspice
¾ c. stone ground whole wheat flour
¼ c. butter OR margarine
¾ c. sugar
2 eggs
⅔ c. milk
½ tsp. vanilla
½ c. chopped nuts

Sift together dry ingredients. Stir in whole wheat flour. Cream together butter and sugar. Add eggs one at a time beating well after each addition. Add dry ingredients alternately with milk and vanilla to creamed mixture. Beat smooth after each addition. Stir in nuts. Turn into greased loaf pan. Bake at 350° for 55 minutes or until done. Cool 10 minutes. Remove from pan.

ANGEL BISCUITS

1 pkg. dry yeast
¼ c. warm water
1½ c. stone ground whole wheat flour
1½ c. all-purpose flour
1 tsp. baking soda
2 tsp. baking powder
2 tblsp. honey
½ c. shortening
1 c. buttermilk

Dissolve yeast in water. Mix dry ingredients. Cut shortening into dry ingredients. Add buttermilk and honey, then yeast. Stir until flour is dampened. Knead on floured board ½ minute. Roll to ½-inch thinckness. Cut with biscuit cutter. Bake at 400° for 12- 15 minutes. No rising is necessary. (Dough can be refrigerated and used as needed for a week.)

BEST WHEAT BISCUITS

2 c. all-purpose flour
2 c. stone ground whole wheat flour
3 tblsp. baking powder
1 tsp. salt
2 tblsp. sugar
1 c. shortening
1⅓ c. milk

Mix together dry ingredients and cut in shortening. Add milk. Knead lightly and roll ½-inch think; cut shapes and bake on ungreased cookie sheet for 10-12 minutes in 450° oven. Serve warm with butter.

SWEET POTATO BISCUITS

¾ c. mashed, cooked sweet potatoes
¼ c. butter OR margarine, melted
2 tblsp. brown sugar
2 tblsp. sugar
1 c. stone ground whole wheat flour
¾ c. all-purpose flour
1½ tblsp. baking powder
1 tsp. baking soda
1 tsp. salt
¾ c. buttermilk

Combine first four ingredients; beat at medium speed of an electric mixer until blended. Combine flour, baking powder, soda and salt in a large bowl, stir well. Add sweet potato mixture and buttermilk to flour mixture; stir just until moistened. Turn dough out onto a floured surface, and knead lightly 6 to 8 times. Roll dough to ½-inch thickness; cut with a 2½-inch biscuit cutter. Place on an ungreased baking sheet; bake at 450° for 18 to 20 minutes. Makes about 10 biscuits.

Patience is the ability to count down before blasting off.

WHOLE WHEAT BISCUITS

1 pkg. dry yeast
2 tblsp. warm water
2½ c. stone ground whole wheat flour
½ c. all-purpose flour
2 tsp. baking powder
½ tsp. salt
1½ tsp. soda
¼ c. margarine, softened
1 tblsp. honey
1 c. buttermilk
Melted butter OR margarine

Dissolve yeast in water; set aside. Combine dry ingredients; cut in margarine until mixture resembles coarse meal. Add yeast mixture, honey, and buttermilk, mixing well. Turn dough out onto a floured surface. Roll dough to ½-inch thickness, then cut into rounds with a 2-inch cutter. Place biscuits on lightly greased baking sheets. Brush with melted butter. Bake at 400° for 12 to 15 minutes. Makes about 1 dozen biscuits.

APPLE MUFFINS

½ c. non-instant dry milk solids
3 tsp. baking powder
½ tsp. salt
½ tsp. allspice
½ tsp. nutmeg
1 tsp. cinnamon
2½ c. stone ground whole wheat flour
1 c. honey
1 c. oil
4 eggs
1 tsp. vanilla
1 c. grated apple
1 c. grated carrot

Preheat oven to 400°. In a large mixing bowl combine milk solids, baking powder, salt, allspice, nutmeg, cinnamon and flour. Combine honey, oil, eggs and vanilla and stir into dry ingredients. Fold in apples and carrots. Spoon into greased muffin tins and bake 15 minutes or until done. Makes about 2 dozen medium-sized muffins.

BANANA CAKE GEMS

1 c. stone ground whole wheat flour
1 c. all-purpose flour
1 tsp. soda
1 tsp. salt
1½ c. sugar
1½ tsp. baking powder, divided
2 large eggs
¾ c. vegetable oil
1 c. mashed banana
½ c. chopped nuts (optional)

Sift together into a medium sized mixing bowl the first five ingredients and 1/4 tsp. baking powder. If using nuts, stir these into the dry ingredients. In a blender or processor, blend the eggs on low; on medium, add the oil, then the bananas and remaining baking powder and blend well. Make a well or "nest" in dry ingredients and pour in banana mixture. Stir just to combine ingredients. Do not mix vigorously. Pour into greased or paper lined medium-sized muffin cups and bake 25 minutes at 350°. These cake-like gems pass as dessert anytime you serve them.

BANANA NUT MUFFINS

1 c. stone ground whole wheat flour
3/4 c. all-purpose flour
3 tsp. baking powder
1/2 tsp. salt
1/2 c. shortening
1 c. sugar
2 eggs
1 1/3 c. mashed bananas
1 c. chopped nuts

Combine flours, baking powder and salt. Cream shortening and sugar until light and fluffy. Beat in eggs, one at a time, blending well after each. By hand, stir in mashed bananas. Add dry ingredients, stirring just enough to moisten. Do not overbeat! Fold in nuts. Pour into 12 greased, 3-inch muffin cups until 3/4 full. Bake at 400° for 20 minutes or until tested done and golden brown.

BEST BLUEBERRY MUFFINS

1 egg
1/4 c. oil
1/2 c. milk
3/4 c. all-purpose flour
3/4 c. stone ground whole wheat flour
1/2 c. sugar
2 tsp. baking powder
1/2 tsp. salt
1 c. blueberries

Grease bottoms of 2 1/2-inch muffin or gem-pan cups. Heat oven to 400°. Beat egg until foamy in a small mixing bowl. Beat in oil and then milk. In separate bowl, stir together dry ingredients. Stir in blueberries. Make a well in the center and pour in liquid ingredients. Stir quickly with fork just until dry ingredients are moistened. Batter will be lumpy. Using 1/4-cup measure (not quite full) dip batter into muffin cups, filling each slightly more than half full. (Try to dip only once for each cup, muffins will be lighter.) Bake 18 to 20 minutes or until golden and muffins test done with pick. Loosen with spatula and turn out. Best served warm. Blueberries may be fresh, frozen or canned (drain). Makes 12 muffins.

BEST EVER PINEAPPLE MUFFINS

1 c. all-purpose flour
1 c. stone ground whole wheat flour
3 tsp. baking powder
½ tsp. salt
1 c. undrained, crushed pineapple
¼ c. sugar
¼ c. margarine OR butter
1 egg

Preheat oven to 400°. Paper line or grease the bottoms of medium sized muffin tins. Stir or sift together the dry ingredients in a separate bowl. In medium mixing bowl cream the sugar and margarine until fluffy. Beat in the egg. Stir in pineapple. Add dry ingredients to the creamed mixture, stirring only enough to moisten the flour. Fill muffin cups ⅔ full. Bake 15-20 minutes. Remove muffins from the tin at once and serve hot. Makes 12 muffins.

BREAKFAST IN A MUFFIN

1 pkg. (8-oz.) cream cheese
4 slices bacon
1 tsp. seasoning salt
¼ c. cheddar cheese
1 egg, beaten
1 c. stone ground whole wheat flour
1 c. all-purpose flour
¼ c. sugar
4 tsp. baking powder
½ tsp. salt
1 c. milk
¼ c. shortening

Cook bacon until crisp. Mix thoroughly creamed cheese, salt, cheddar cheese and egg. Crumble bacon and stir into cheese mixture. Combine flours, sugar, baking powder and salt. Add milk, egg, and shortening; beat until fairly smooth, about 1 minute. Divide batter among 18 cupcake liners placed in muffin tins. With a teaspoon, place the cheese mixture on top of each muffin, dividing it among the muffins and pressing cheese mixture down into batter. Bake in preheated 425° oven 15 to 20 minutes. Makes 18.

CARROT MUFFINS

1 c. milk
¼ c. molasses
¼ c. oil
2 eggs
1 tsp. vanilla
2 c. stone ground whole wheat flour
¼ c. all-purpose flour
2 tsp. baking powder
½ tsp. soda
½ tsp. salt
½ tsp. nutmeg
¾ c. grated raw carrots

In medium sized bowl, combine milk, molasses, oil, eggs and vanilla. In a large bowl, stir together flour, baking powder, soda, salt and nutmeg. Make a well in center of dry ingredients and stir in liquids, only until blended. Batter will be lumpy—don't overmix. Add carrots. Fill muffin tins ⅔ full with batter. Bake in 350° oven about 20 to 25 minutes. Makes 18 muffins.

CHEESE MUFFINS

1 c. stone ground whole wheat flour
¾ c. all-purpose flour
3 tblsp. sugar
1 tblsp. baking powder
¾ tsp. garlic salt
Dash of cayenne pepper
1 c. shredded cheddar cheese
½ tsp. grated lemon peel
1 egg, slightly beaten
1 c. milk
¼ c. butter, melted

Combine flour, sugar, baking powder, garlic salt and cayenne. Reserve 2 tblsp. cheese for topping. Combine remaining cheese with lemon peel and flour mixture. Combine beaten egg with milk and melted butter. Pour all at once into dry ingredients, stirring until just moistened. Divide equally into nine greased, 3-inch muffin tins. Sprinkle tops with reserved cheese. Bake at 425° for 20 minutes.

CINNAMON GRAHAM MUFFINS

1 egg, beaten
¾ c. buttermilk
¼ c. oil
1 tsp. vanilla
1½ c. cinnamon graham cracker crumbs
½ c. raisins OR dates
1 c. stone ground whole wheat flour
2 tblsp. sugar
1½ tsp. baking powder
1 tsp. baking soda
¼ tsp. salt

Combine egg, buttermilk, oil, and vanilla. Stir in cracker crumbs and raisins or dates; set aside. Combine flour, sugar, baking powder, baking soda and salt. Add to egg mixture, stirring just to moisten. Fill greased muffin pans ⅔ full. Bake at 400° for 20 to 25 minutes. Yield: 12 muffins.

CRACKED WHEAT MUFFINS

1 c. cracked wheat cereal
1 c. milk
1 large OR 2 small eggs
¼ c. vegetable oil
1¼ c. all-purpose flour
1 tblsp. baking powder
¼ tsp. salt
½ c. granulated sugar

Preheat oven to 375°. Combine cracked wheat and milk, stir to combine and let stand 3 minutes until cereal is softened. Add egg and oil and beat well. Sift remaining ingredients together into egg mixture and stir only until combined. Portion evenly into 12 greased muffin cups. Bake about 25 minutes until golden brown. Makes 12 muffins.

MIGHTY MUFFINS

1½ c. whole bran cereal
1 c. skim milk
1 egg, beaten
⅓ c. molasses
¼ c. butter OR margarine (melted)
½ c. all-purpose flour
½ c. stone ground whole wheat flour
2 tblsp. toasted wheat germ
2 tsp. baking powder
½ tsp. baking soda
½ c. raisins
½ c. chopped nuts

Combine bran and milk, let stand 3 minutes. Stir in egg, molasses and butter; set aside. Combine flours, wheat germ, baking powder and soda. Make well in center. Add bran mixture; stir until moistened. Fold in raisins and nuts. Grease muffin tins or line with paper baking cups. Fill ⅔ full. Bake in 400° oven 20-25 minutes. Cool. Store, covered, in refrigerator. Makes 12 muffins.

OATMEAL MUFFINS

1 c. stone ground whole wheat flour
¼ c. brown sugar
3 tsp. baking powder
½ tsp. salt
6 tblsp. shortening
1 c. uncooked oatmeal
1 beaten egg
1 c. milk

Sift flour, sugar, baking powder and salt together. Cut in shortening. Add oatmeal, then egg and milk. Stir only until mixed. Pour into 12 muffin cups. Sprinkle with a mixture of ⅓ c. brown sugar, 1 tblsp. flour, 2 tsp. cinnamon and 1 tblsp. melted butter. Bake at 425° for 15 to 20 minutes.

SPICE MUFFINS

1 c. butter OR margarine, softened
2 c. sugar
2 eggs
2 c. applesauce
3 tsp. cinnamon
2 tsp. allspice
½ tsp. cloves
2 tsp. baking powder
1 tsp. salt
2 tsp. baking soda
2 c. stone ground whole wheat flour
1¾ c. all-purpose flour
1 c. chopped pecans
Confectioners' sugar

Preheat oven to 350°. Cream butter and sugar with electric mixer. Add eggs, one at a time. Mix in applesauce and spices. Sift together salt, baking powder, soda and flour. Add to batter and beat well. Stir in nuts. Bake in lightly greased muffin tins for 15 to 20 minutes. Sprinkly with confectioners' sugar. Covered batter will keep in the refrigerator for 3 weeks. Makes 2 dozen regular or 6 dozen small muffins.

STRAWBERRY RHUBARB MUFFINS

1 c. stone ground whole wheat flour
¾ c. all-purpose flour
½ c. sugar
2½ tsp. baking powder
¾ tsp. salt
1 egg, lightly beaten
¾ c. milk
⅓ c. vegetable oil
¾ c. minced, fresh rhubarb
½ c. sliced strawberries
6 small strawberries, halved
Sugar

Combine flour, sugar, baking powder and salt in large bowl. Combine egg, milk and oil in small bowl; stir into flour mixture with fork until just moistened. Fold rhubarb and sliced berries into batter. Fill well-greased or paper-lined muffin tins ⅔ full with batter. Press a strawberry half gently into top of each muffin; sprinkly tops generously with sugar. Bake at 400° for 20 to 25 minutes until golden.

WHOLE WHEAT MUFFINS

1½ c. stone ground whole wheat flour
¼ c. honey
1 tsp. salt
4 tsp. baking powder
1 egg, well beaten
1 c. milk
3 tblsp. cooking oil

Combine dry ingredients and honey in a large bowl. Set aside. Mix together egg, milk and oil. Add the liquid ingredients all at once to the dry ingredients. Stir just until flour mixture is moistened. Fill greased muffin pans ⅔ full and bake in 425° oven for 15 minutes or until done. Makes 10-12.

WHOLE WHEAT CORNMEAL MUFFINS

1 c. stone ground whole wheat flour
1 c. cornmeal
4 tsp. baking powder
2 beaten eggs
1⅓ c. milk
¼ c. honey
¼ c. oil
½ tsp. salt
Few sesame seeds

In bowl, stir flour, cornmeal, baking powder and salt together. Set aside. In separate bowl combine eggs, milk, honey and oil. Add all at once to dry mixture. Stir until moist (thin batter). Spoon into well greased muffin tins. Sprinkle with seeds. Bake at 400° for 18-20 minutes. Makes 18 muffins.

WHOLE WHEAT PUMPKIN MUFFINS

½ c. all-purpose flour
¾ c. stone ground whole wheat flour
½ c. toasted wheat germ
2 tblsp. sugar
3 tsp. baking powder
½ tsp. salt
½ tsp. cinnamon
½ tsp. nutmeg
2 egg whites
¾ c. skim milk
½ c. canned pumpkin
¼ c. cooking oil
1 tsp. vanilla

Stir together the flours, wheat germ, sugar, baking powder, salt, cinnamon and nutmeg. Make a well in center. Combine egg whites, milk, pumpkin, oil and vanilla; add all at once to dry ingredients. Stir just till moistened. Spoon into greased muffin cups. Bake in 400° oven 20-25 minutes. Cool 2 minutes in pan. Remove; cool. Makes 12.

BEST-YET COFFEE CAKE

1¼ c. stone ground whole wheat flour
1 c. all-purpose flour
¼ c. cornstarch
½ tsp. salt
1 tsp. cinnamon
¾ c. white sugar
1 c. brown sugar
¾ c. salad oil
1 egg
1 tsp. soda
1 tsp. baking powder
¾ c. buttermilk

Combine first eight ingredients. Mix to fine crumbs and divide equally into two mixing bowls. To one mixture add remaining ingredients, but only ½ cup of the buttermilk. Beat for two minutes, add remaining ¼ cup of buttermilk and beat two minutes more. Pour into 9 x 13-inch pan. Sprinkle with second half of crumb mixture. Bake at 350° for 30 minutes.

BLUEBERRY BUCKLE

¾ c. sugar
¼ c. margarine
1 egg, beaten
1 tsp. lemon extract
½ c. milk
1 c. stone ground whole wheat flour
¾ c. all-purpose flour
¾ tsp. salt (optional)
2½ tsp. baking powder
2 c. blueberries, fresh, frozen and thawed OR canned and drained

Cream sugar and margarine; beat in egg, lemon extract and milk. Stir in baking powder and salt, if used, then stir in flours. Gently fold in blueberries. Batter will be thick. Turn into a greased 9 x 13 x 2-inch pan. To make streusel topping, combine ½ cup sugar, ⅓ c. flour and 1 tsp. cinnamon; cut in ¼ cup margarine until mixture is crumbly. Sprinkle over batter. Bake in a preheated 350° oven for 35 minutes. Makes 12 servings.

COFFEE CAKE BARK

1 c. raisins
⅔ c. hot coffee
⅔ c. shortening
1 c. sugar
2 eggs
1¼ c. stone ground whole wheat flour
1 tsp. baking powder
1 tsp. soda
¼ tsp. salt
½ tsp. cinnamon
½ c. pecans
2 tblsp. milk
4 tblsp. margarine
2 c. powdered sugar
¾ tsp. vanilla
½ tsp. almond extract

Pour hot coffee over raisins. Let stand until cool. Cream shortening and sugar. Add eggs. Stir dry ingredients together. Add with pecans, then stir in cooled raisins and coffee. Spread on greased 10 x 15-inch cookie sheet. Bake for 20 minutes at 350°. Frost with icing. For icing, heat milk and margarine. Add powdered sugar, vanilla and almond extract.

CRUNCHY PEACH COFFEE CAKE

Topping:
½ c. stone ground whole wheat flour
½ c. firmly-packed brown sugar
¼ c. wheat germ
1 tsp. cinnamon
¼ c. butter OR margarine
1 can (1 lb. 13 oz.) sliced cling peaches

Batter:
1 c. stone ground whole wheat flour
½ c. all-purpose flour
¼ c. wheat germ
¼ c. sugar
2 tsp. baking powder
½ tsp. salt
¼ c. butter OR margarine
2 eggs, beaten
½ c. milk

Prepare topping; combine ½ cup flour, brown sugar, ¼ c. wheat germ and cinnamon in bowl; mix well. Cut in ¼ cup butter OR margarine until mixture is crumbly, using a pastry blender or 2 knives. Set aside. Drain sliced peaches. Prepare batter: combine 1½ cups flours, ¼ cup wheat germ, sugar, baking powder and salt in bowl; mix well. Cut in ¼ cup butter OR margarine until mixture is crumbly using a pastry blender or 2 knives. Add eggs and milk; stir just until dry ingredients are moistened. Spread batter over bottom of greased 9 x 13 x 2-inch baking pan. Sprinkle with ½ the reserved topping mixture. Arrange peach slices in rows on top. Sprinkle with remaining topping mixture. Bake at 350° for about 25 minutes or until done. Serve warm or cold.

FRESH APPLE COFFEE CAKE

1½ c. stone ground whole wheat flour
4 tblsp. cornstarch
1 c. quick cooking oats
¼ c. wheat germ
1 tblsp. baking powder
¾ tsp. cinnamon
¼ tsp. nutmeg
¼ tsp. salt
½ c. honey
½ c. vegetable oil
⅓ c. milk
2 eggs, beaten
3 c. finely chopped apples

Mix dry ingredients together well. Mix liquid ingredients together. Add dry ingredients to liquid ingredients until moistened. Add apples. Bake at 350° in 9-inch square baking pan for 30-35 minutes.

FRESH STRAWBERRY COFFEE CAKE

½ c. sugar
1 c. all-purpose flour
2 tsp. baking powder
½ tsp. salt
½ c. milk
1 egg
2 tblsp. melted butter
1½ c. fresh strawberries, sliced
Topping:
½ c. stone ground whole wheat flour
½ c. sugar
¼ c. butter
¼ c. chopped walnuts

Combine batter ingredients, except strawberries; beat for 2 minutes to blend. Spread into greased 8 x 8 x 2-inch pan. Sprinkle berries evenly over batter. Combine topping ingredients, mix to crumbs. Sprinkly over strawberries. Bake at 375° for 35-40 minutes.

GLAZED APPLE COFFEE CROWN

2 c. stone ground whole wheat flour
2-2½ c. all-purpose flour
⅓ c. sugar
1 tsp. salt
1 pkg. dry yeast
1 c. milk
½ c. water
¼ c. butter OR margarine
1 egg

Filling:
¾ c. sugar
1 tsp. cinnamon
¼ c. butter OR margarine, softened
1 3-oz. pkg. cream cheese, softened
2 c. peeled, finely chopped apples
⅓ c. firmly packed brown sugar
½ tsp. cinnamon

Glaze:
1 c. powdered sugar
1 tblsp. milk
1 tblsp. butter OR margarine, softened
1 tsp. lemon juice

Grease 12-cup fluted tube pan or 10-inch tube pan. Lightly spoon flour into measuring cup, level off. In large bowl, combine 2 c. whole wheat flour, 1/3 c. sugar, salt and yeast. In small saucepan, heat milk, water and margarine until very warm. Add warm liquid and egg to flour mixture. Blend at low speed until moistened; beat 2 minutes at medium speed. By hand, stir in 2 cups flour. On floured surface, knead in 1/2 to 1 cup flour until smooth and elastic, about 5 to 8 minutes. Place in greased bowl; cover loosely with plastic wrap and cloth towel. Let rise in warm place until light and doubled in size, about 1 hour. In small bowl, combine 3/4 c. sugar, 1 tsp. cinnamon, 1/4 c. margarine and cream cheese; blend until smooth. In small bowl, combine apples, brown sugar and 1/2 tsp. cinnamon. Divide dough in half. On lightly floured surface, roll out half into 18 x 8-inch rectangle. Spread with half of cream cheese mixture to within 1/2 inch of edge, then spread with half of apples. Starting at longer side, roll up tightly. Pinch edges and ends to seal. Repeat with remaining half of dough. Place both rolls in prepared pan. Cover; let rise in warm place until light and doubled in size, about 1 hour. Heat oven to 350°. Bake 45 to 55 minutes or until golden brown. Remove from pan immediately. In small bowl, combine all glaze ingredients; drizzle over warm coffee cake. Makes 16 servings.

OVERNIGHT COFFEE CAKE

2/3 c. margarine
1 c. sugar
1/2 c. brown sugar
2 eggs, beaten
1 c. stone ground whole wheat flour
3/4 c. all-purpose flour
1/4 c. cornstarch
2 tblsp. powdered milk
1 c. buttermilk OR sour milk
1 1/2 tsp. baking soda
1 1/2 tsp. baking powder
1/2 tsp. salt
1 tsp. cinnamon
Topping:
1/2 c. brown sugar
1 tsp. cinnamon
1/2 tsp. nutmeg
1/2 c. chopped nuts

Cream together the margarine and sugars. Add eggs, mixing thoroughly. Combine dry ingredients. Add alternately to first mixture with buttermilk, starting and ending with dry ingredients. When mixed thoroughly, place in 9 x 13-inch baking dish and cover with topping. Topping: Combine all ingredients and sprinkle on top of coffee cake. Cover baking dish with aluminum foil or plastic wrap and refrigerate overnight. Bake at 350° for 30-35 minutes.

SUPERB BRUNCH COFFEE CAKE

½ c. butter OR margarine
¾ c. sugar
1 tsp. vanilla
3 eggs
1½ c. raisins
1¾ c. stone ground whole wheat flour
¼ c. cornstarch
1½ tsp. baking powder
1½ tsp. soda
1 tsp. salt
1 c. sour cream
Pecan Praline Mix

Beat butter, sugar and vanilla until fluffy. Blend in eggs 1 at a time, then raisins. Add flour (sifted with baking powder, soda and salt) alternately with sour cream; mix until smooth. Spread half the batter in greased, floured 10-inch tube pan. Sprinkle with half the pecan praline mix. Repeat layers. Bake at 350° for 50 minutes or until done. Cool in pan 10 minutes; turn out on wire rack. Makes 1 10-inch coffee cake. Pecan Praline Mix: 1 c. firmly packed brown sugar; 2 tsp. cinnamon; ⅓ c. butter OR margarine, softened; ¾ c. coarsely chopped pecans. Mix brown sugar and cinnamon together, cut in butter until crumbly. Add pecans.

BEST BAKED DO-NUTS

⅔ c. shortening
1 c. sugar
2 eggs
1 c. milk
1 c. stone ground whole wheat flour
2 c. all-purpose flour
4½ tsp. baking powder
½ tsp. salt
½ tsp. nutmeg
Topping:
⅓ c. butter OR margarine, melted
½ c. sugar
½ tsp. cinnamon

Preheat oven to 350°. Blend shortening with sugar. Add eggs and beat well. Combine or sift dry ingredients together. Add alternately with milk. Fill greased muffin pans ⅔ full. Bake about 15 minutes or until lightly brown. Prepare topping. Dip warm donuts in melted margarine and roll in cinnamon-sugar mixture. Makes 2 dozen medium do-nuts.

NO-FRY RAISED DOUGHNUTS

2 pkg. dry yeast
¼ c. warm water
1½ c. milk, scalded & cooled
½ c. sugar
1 tsp. salt
1 tsp. nutmeg
½ tsp. cinnamon
2 eggs
⅓ c. shortening, melted
2½ c. stone ground whole wheat flour
2 c. unbleached flour

In a large mixing bowl, dissolve yeast in warm water; add milk, sugar, salt, nutmeg, cinnamon, eggs and shortening. Add whole wheat flour, blend on low speed of mixer, scraping bowl occasionally. Mix on medium speed of mixer 10 minutes. Stir in remaining flour until dough is smooth. Cover; let rise until double. Turn onto well-floured board and roll around to coat with flour; the dough will be soft to handle. Roll dough about ½-inch thick; cut with a doughnut cutter. Place 2 inches apart on greased baking sheet. Brush with melted butter; cover and let rise until double. Bake in preheated oven, 425°, 8 to 10 minutes. Immediately brush with melted butter and sprinkle with sugar or glaze.

WHOLE WHEAT DOUGHNUTS

2½ c. all-purpose flour
1 tblsp. baking powder
½ tsp. baking soda
1 tsp. salt
½ tsp. nutmeg
¼ tsp. cinnamon
1¾ c. stone ground whole wheat flour
2 beaten eggs
½ c. white sugar
½ c. brown sugar
3 tblsp. melted shortening
2 tblsp. grated lemon rind
½ tsp. vanilla
1 c. buttermilk OR sour milk

Sift together flour, baking powder, baking soda, salt, nutmeg and cinnamon. Add sugars to eggs and beat well. Add shortening, lemon rind and vanilla. Mix well. Add flour mixture alternately with buttermilk or sour milk to egg mixture. Mix lightly to make a soft dough. Do not add too much flour. Turn out on lightly floured surface. Cut with doughnut cutter. Fry in deep hot fat at 365°, turning frequently until brown. Drain thoroughly on absorbent paper. Dust with cinnamon and sugar if desired. Makes 2 dozen 2½-inch doughnuts.

WHOLE WHEAT POTATO DOUGHNUTS

2 c. scalded milk
1 c. butter OR margarine
1 tsp. mace
1 c. mashed potatoes
1 c. sugar
3 eggs, beaten
2 pkg. yeast
2 tsp. salt
6 c. stone ground whole wheat flour
4-6 c. unbleached flour
1 c. potato water

Dissolve yeast in 1 cup potato water. Mix milk, butter, mace, mashed potatoes, and sugar. Set aside. Mix beaten eggs and salt and add to yeast. Add yeast mixture to the milk mixture. Blend in the whole wheat flour, mix 5 minutes and let set for 15 minutes. Add unbleached flour. Dough will be sticky. Chill overnight. Let rise, shape into doughnuts, raise and fry in deep hot fat, 365°, turning to brown both sides. Roll in sugar or glaze: 1 lb. powdered sugar, 6 tblsp. water, 1 tblsp. vanilla.

CHOCOLATE GLAZE

4 oz. semi-sweet chocolate
1/3 c. butter OR margarine
2 c. confectioners' sugar
1½ tsp. vanilla
4 to 6 tblsp. water, hot

Over low heat, melt chocolate and butter. Remove from heat and stir in confectioners' sugar, vanilla, and 4 tblsp. hot water. Add additional water if necessary for proper consistency.

CREAMY GLAZE FOR DOUGHNUTS

1/3 c. butter OR margarine
2 c. confectioners' sugar
1½ tsp. vanilla
4 to 6 tblsp. water

Melt butter and blend in remaining ingredients, using 4 tblsp. water. Add additional water until glaze is of proper consistency.

BLUEBERRY GRIDDLE CAKES

1 well beaten egg
1 c. milk
1/4 c. butter OR margarine, melted
1/2 c. stone ground whole wheat flour
1/2 c. all-purpose flour
3 tsp. baking powder
2 tblsp. sugar
3/4 tsp. salt
1 c. drained canned, frozen, or fresh blueberries

Combine egg, milk and butter. Sift the dry ingredients, gradually add to liquid using rotary beater. Drop batter on hot lightly greased griddle, using 1/3 cup measure. Sprinkle about 2 tblsp. of blueberries over each cake. When underside is golden, turn and brown other side.

PORK AND ZUCCHINI PANCAKES

1 lb. ground pork
1 tsp. salt
1/4 tsp. thyme leaves
1/8 tsp. pepper
1 egg, beaten
1/3 c. milk
1 tblsp. oil
1/2 c. stone ground whole wheat flour
2 tsp. baking powder
1/4 tsp. salt
1/2 c. shredded zucchini
1/4 c. plain yogurt
2 tblsp. cooking fat

Brown ground pork in large frying pan; pour off drippings. Sprinkly 1 teaspoon salt, thyme and pepper over meat, stirring to combine. Combine egg, milk and oil in mixing bowl; stir in flour, baking powder and salt just until lumps disappear. Stir in zucchini and half of ground pork mixture. Keep remaining pork warm. Place 2 tblsp. cooking fat on griddle or in large frying pan and heat. Divide pork-zucchini mixture into four equal portions. Cook pancakes over moderate heat 4 to 5 minutes on each side. Top each pancake with 1 tblsp. yogurt and remaining pork. Makes 4 servings.

WHOLE WHEAT PANCAKES

2 c. stone ground whole wheat flour
1/4 c. cornstarch
1 tsp. salt
2 tblsp. baking powder
1/4 c. cooking oil
2 tblsp. honey (1/8 cup)
2 eggs, medium to large
2 c. milk OR buttermilk

In large bowl, combine dry ingredients. Add the rest of the ingredients and mix well. If batter is too thick add a little more milk. Cook pancakes on hot, lightly greased griddle.

WHOLE WHEAT PANCAKES

1 1/4 c. stone ground whole wheat flour
3/4 c. all-purpose flour
2 tblsp. brown sugar, packed
1 tblsp. baking powder
1/2 tsp. salt
2 beaten eggs
1 1/2 c. milk
3 tblsp. cooking oil or melted shortening

Stir together dry ingredients. Beat together eggs, milk and oil or melted shortening; add to dry ingredients, beating till blended. Bake on hot, lightly greased griddle, using 2 tblsp. batter for each pancake.

WHOLE WHEAT WAFFLES

2 c. stone ground whole wheat flour
1/4 c. cornstarch
4 tsp. baking powder
1 tsp. soda
1/2 tsp. salt
3 beaten egg yolks
2 c. buttermilk
1 tblsp. sugar
1/2 c. vegetable oil
3 stiffly beaten egg whites

Stir dry ingredients together. Mix beaten egg yolks, milk and oil with mixer. Add dry ingredients and mix well. Fold in egg whites. Cook in waffle iron until brown on both sides.

WHOLE-WHEAT CREPES

1½ c. milk
1 c. stone ground whole wheat flour
2 eggs
1 tblsp. cooking oil
¼ tsp. salt

In bowl, combine milk, flour, eggs, cooking oil and salt; beat with rotary beater until blended. Heat lightly greased 6-inch skillet. Remove from heat. Spoon in about 2 tblsp. batter; lift and tilt skillet to spread batter. Return to heat; brown on one side. (Or cook on inverted crepe pan). Invert pan over paper toweling; remove crepe. Repeat with remaining batter to make 16 to 18 crepes, greasing skillet occasionally.

SHRIMP CREPES

1 pkg. (10 oz.) frozen asparagus spears
1 can (4½ oz.) medium shrimp
1 can (6½ or 7 oz.) tuna, drained and flaked
⅓ c. mayonnaise OR salad dressing
¼ c. finely chopped celery
1 tblsp. finely chopped pimento
3 egg yolks
½ tsp. salt
Dash pepper
¼ c. mayonnaise OR salad dressing
3 egg whites
6 whole wheat crepes (recipe above)
¾ c. shredded Swiss cheese (3 oz.)

In saucepan, cook asparagus according to package directions; drain well. Set aside 6 shrimp. In small bowl, combine remaining shrimp, ⅓ c. mayonnaise, celery and pimento. In mixing bowl, beat egg yolks, salt and pepper until thick and lemon-colored; stir in ¼ c. mayonnaise. In mixing bowl, beat egg whites with electric mixer until stiff peaks form. Carefully fold egg yolk mixture into beaten egg white. Place crepes, browned side up and equal distance apart on greased baking sheet. Spread each crepe with about 3 tblsp. shrimp filling. Sprinkle with cheese. Place asparagus spears atop. Spoon egg mixture over the asparagus. Bake at 350° until topping is golden brown and filling is warm, about 15 mintues. Use wide spatula to transfer crepes to individual service plates. Garnish with reserved shrimp. Serve immediately. Makes 6 crepes.

MAIN DISHES

APPLE-CHEDDAR QUICHE

¼ c. butter OR margarine
1 tblsp. finely chopped onion
1½ c. crushed stone-ground whole wheat crackers
¼ c. finely chopped walnuts
2 large tart apples, peeled, cored, and sliced
3 eggs
1 c. cream-style cottage cheese
1 c. shredded cheddar cheese (4 oz.)
¼ c. milk
½ tsp. salt
Dash pepper
Ground nutmeg
Shredded Cheddar cheese (optional)

In 4-cup glass measure, combine butter or margarine and onions. In microwave oven cook, covered, on 100% (high) for 1 to 2 minutes or until onion is tender. Stir in crushed crackers and walnuts. Press mixture into a 9-inch microwave oven-safe pie plate, forming crust. Microwave, uncovered, on high for 2 to 3 minutes or until set, giving dish half-turn once or twice during cooking. In microwave oven-safe mixing bowl, microwave apple slices in small amount of water, covered, on high for 2 to 3 minutes or until tender; drain. Arrange apple slices in crust. For filling, in blender container or food processor bowl combine eggs, cottage cheese, 1 cup cheddar cheese, milk, salt and pepper. Cover and blend until mixture is nearly smooth. Pour into crust. Sprinkle nutmeg atop. In conventional oven, bake at 325° about 45 minutes or until knife inserted in center comes out clean. Sprinkle with additional cheddar cheese if desired. Let stand 10 minutes. Makes 5 servings.

APPLE MEAT LOAF

1½ lb. ground beef
2 eggs
¼ c. milk
¾ c. whole wheat bread crumbs
2 apples, chopped
¼ c. finely chopped onion
⅛ tsp. pepper
⅛ tsp. nutmeg

Preheat oven to 350°. Combine ingredients well and bake for 1 hour.

Old age is about the only thing that comes to us without effort.

BOUNTIFUL WHEAT CASSEROLE

1 lb. lean ground beef
16 oz. can tomatoes
1 tblsp. salt
1 tsp. each oregano, basil and garlic powder
3 c. cooked wheat kernels
4 c. cabbage, coarsely-shredded
½ c. sour cream
1 c. **each** finely chopped onions and green peppers
1 c. grated mozzarella cheese

Saute beef until lightly browned. Stir in tomatoes, seasonings, wheat and cabbage. Cover and cook 10 to 15 minutes, or until cabbage is tender crisp. Stir in sour cream, onions and green peppers. Note: Saute onion and green pepper with beef, if desired. Heat. Spoon into warm serving dish and sprinkle with cheese. Let stand a few minutes before serving to allow cheese to melt. Makes 6 to 8 servings.

CABBAGE BURGERS

Cabbage Filling:

1 medium head cabbage, shredded
1 medium onion, chopped
1½ lb. ground beef
3 tsp. salt
½ tsp. pepper

Mix in large kettle and steam over low heat until done. Remove from heat; cool.

Roll Mixture:

1 pkg. dry yeast
¼ c. warm water
2 c. scalded milk
2 tblsp. sugar
2 tsp. salt
1 tsp. lard OR shortening
1-2 c. all-purpose flour
2 c. stone ground whole wheat flour

Soften yeast in warm water. Combine milk, sugar, salt and lard. Add 2 cups of flour to make moderately stiff dough. Turn out on lightly floured board. Knead until smooth (about 8 minutes). Put into greased bowl and let rise until double in bulk (45 minutes). Cut dough into balls and roll out to size of saucer. Place 2 tblsp. of Cabbage Filling onto center of rolled out dough. Bring sides up and pinch together to seal. Put into greased pan and let rise 10 minutes. Bake 20 minutes in 350° oven. Can be frozen in foil for two weeks. Just reheat when ready to eat.

CHEESY SPINACH QUICHE

2 c. chopped fresh spinach, firmly packed
6 eggs
¼ c. stone ground whole wheat flour
1 tsp. dry mustard
1 tsp. lemon juice
½ tsp. salt
½ tsp. baking powder
⅔ c. chopped onion
6 tblsp. butter
1 c. chopped ham
1 c. grated Swiss cheese, firmly packed
1 c. grated Cheddar cheese, firmly packed

In glass dish, combine spinach, onion, and butter; microwave on high 2½ minutes. Set aside. Beat eggs; add and beat well flour, mustard, lemon juice, salt and baking powder. Gently fold in ham, cheeses, and spinach mixture. Pour mixture into 10½-inch quiche pan. Cover with plastic wrap and microwave on high for three minutes. Remove plastic wrap and gently push outside edges of quiche to center of dish. Microwave on medium-high for 10 minutes rotating dish twice. Let stand for 5 minutes. Makes 6 servings. This quiche makes its own crust.

CHICKEN AND SAUSAGE STEW

1 whole chicken
Cold Water
Bay leaf, black pepper, celery leaves, garlic, onion and thyme for seasoning stock
1½ c. large-diced celery
1½ c. large-diced white OR yellow onions
1½ c. large-diced green AND/OR red bell pepper
8 oz. cooking oil, fat or drippings
8 oz. all-purpose flour
3 c. beef stock, heated to boiling
Basil leaves, thyme leaves, allspice, ground cloves, poultry seasoning, cayenne pepper, liquid hot pepper sauce, Worchestershire sauce and salt to taste for seasoning.
1½ lbs. smoked sausage such as hot links or kielbasa (Polish), sliced on diagonal
Hot cooked wheat
1 bunch green onions, sliced
Liquid hot pepper sauce or hot red pepper flakes

In large cooking pot, cover chicken with cold water. Add bay leaf, black pepper, celery leaves, garlic, onion and thyme to season. Bring to boil slowly, reduce heat to simmer and cook until chicken is tender. Immediately remove chicken to colander and cool with cold running water. Reserve stock for another use. Remove all bones, fat and skin from meat; chop coarse and set aside. While chicken is cooking, chop vegetables and prepare brown roux. For roux, combine oil, fat or drippings and flour in heavy skillet. Cook over medium heat, stirring constantly, until mixture is a golden-brown color. To stop browning, remove roux from skillet into heatproof container. In skillet, in a small amount of oil, fat or drippings, lightly saute chopped onions, celery, and green pepper. Heat beef stock in large pot; add sauted vegetables. In skillet in which vegetables were cooked, brown sausage. Add sausage and chicken pieces to beef stock. Bring mixture to boiling. Add brown roux slowly, stirring and cooking until sauce is desired consistency. Season stew heavily with thyme and basil; lightly season with allspice, cloves, poultry seasoning, red pepper, liquid hot pepper sauce, Worchestershire sauce and salt. Cook over low heat to blend flavors. Serve over hot cooked wheat, topped with sliced green onions. Pass extra liquid hot sauce or dried red pepper flakes, if desired. Makes about 4 servings.

CHICKEN AND WHEAT BAKE

½ c. chopped onion
2 tblsp. oil
1½ c. wheat kernels
2-3 tblsp. curry powder
1 16-oz. pkg. mixed vegetables
2 c. cubed, cooked chicken
2-3 c. chicken broth

In large saucepan, cook onion in hot oil until tender. Add wheat kernels, curry powder and chicken broth. Bring to a boil; reduce heat and cook, covered for 50 minutes. Stir in vegetables and chicken. Turn into 2 qt. buttered casserole. Bake, covered, at 350° for 30-35 minutes. Uncover, and bake 5 minutes more.

CHICKEN & WHEAT CASSEROLE

2 c. cooked wheat kernels
1 tsp. salt
2 c. chicken, cooked & diced
1 small can mushrooms
½ c. almonds, slivered
1 c. celery, diced
2½ tblsp. flour
2 tblsp. instant chicken bouillon
3 c. chicken broth
½ c. Chinese noodles

Combine wheat with chicken, mushrooms, salt, almonds and celery. Add flour and bouillon to chicken broth and pour over other ingredients. Bake for one hour at 350°. Before serving top with noodles.

CHICKEN NOODLE BAKE

⅓ lb. noodles (2 cups) (whole wheat noodles are very good)
1½ tsp. salt
1½ qt. water
1 c. cooked wheat kernels
1 c. cooked, chopped chicken
10½ oz. can mushroom soup
¼ c. finely chopped onion (optional)
1 c. chicken broth
¾ c. crushed potato chips

Cook noodles in salt water. Drain. Mix wheat, chicken, mushroom soup, onion and broth with noodles. Turn into greased 1½ quart baking dish. Sprinkle with crushed potato chips. Bake 35 minutes at 350°.

CHINESE WHEAT CASSEROLE

1 lb. ground beef
1 onion, chopped
1 c. chopped celery
1 10¾-oz. can cream of mushroom soup
1 10¾-oz. can cream of chicken soup
1 16-oz. can Chinese vegetables, drained
½ c. water
1 tblsp. soy sauce
1 c. cooked wheat kernels
1 small can Chinese noodles

Brown meat with onions. Mix well with all other ingredients except noodles. Bake 1 hour in 350° oven. Top with Chinese noodles and bake 15 minutes more. Makes 6 to 8 servings.

CRACKED WHEAT-GROUND BEEF CASSEROLE

1 lb. lean ground beef
2 stalks celery, chopped (1 cup)
1 large green pepper, chopped
1 medium onion, chopped
1 clove garlic, minced
1½ tsp. salt
⅛ tsp. pepper
1 16-oz. can tomatoes, cut up
1 c. cracked wheat
1 c. water
½ c. raisins
⅓ c. shelled sunflower seeds
Cheddar cheese, sliced and halved diagonally (optional)

In skillet cook ground beef, celery, green pepper, onion, garlic, salt and pepper till meat is browned and vegetables are crisp-tender; drain off excess fat. Stir in undrained tomatoes, cracked wheat, water, raisins and sunflower seeds. Turn mixture into a 2-quart au gratin dish or casserole. Bake, covered, in 375° oven for about 35 minutes or till wheat is tender and mixture is heated through. If desired, uncover and top with a few half slices of cheese during the last 5 minutes of baking. Serves 6.

DINNER IN A SKILLET

½ lb. ground chuck
1 egg, beaten
¼ c. milk
¼ c. whole wheat bread crumbs
1½ tblsp. onion, chopped
½ tsp. salt
¼ tsp. dry mustard
2 tblsp. whole wheat flour
2 tblsp. oil
1 can condensed tomato or mushroom soup diluted with
¾ c. milk
1½ c. cooked vegetables or 1 pkg. mixed frozen vegetables, thawed
½ tsp. salt

Combine chuck, egg, milk, crumbs, onion, salt & mustard. Shape into 12 1-inch balls. Sprinkle with flour. Brown balls in oil on all sides. Arrange on sides of skillet. Into center, gradually pour diluted soup. Top with vegetables and salt. Cover. Simmer 10-12 minutes. (If using frozen vegetables, cook mixture 25 minutes).

FARM-STYLE BRUNCH SAUSAGE SOUFFLE

1-1½ lb. sausage links
8 slices whole wheat bread, cubed
2 c. grated cheddar cheese
4 eggs
1 can cream of mushroom soup
3 c. milk
¾ tsp. dry mustard

Brown links and cut into thirds. Place bread cubes in the bottom of an 8 x 12-inch dish. Top with cheese. Put sausage on top of cheese. Beat eggs with 2½ cups of milk and mustard and pour over sausage. Refrigerate overnight. Next day, dilute soup with ½ cup of milk and pour over top. Bake uncovered at 300° for 1½ hours, or until set. Serves 6 to 8.

GOLDEN PUFF

10 slices whole wheat bread
6-8 eggs
3 c. milk
2 c. diced ham (or bacon or shrimp)
½ tsp. salt
¾ tsp. dry mustard
2 c. diced cheese
2 tblsp. fresh parsley or 1 tblsp. dried parsley

Remove crusts from bread. Cube bread into small pieces. Beat eggs, milk and seasonings. Add all other ingredients. Put in a buttered 11½ x 7½-inch dish. Bake 1 hour at 325°. Serves 8 to 12.

GROUND BEEF-MUSHROOM ROLL

1 4-oz. can sliced mushrooms
1 lb. lean ground beef
½ c. fine, dry whole wheat breadcrumbs
1 egg, beaten
½ tsp. salt
¼ tsp. garlic powder
4 slices bacon, cooked and crumbled
1 small onion, chopped
1 c. (4 oz.) shredded mozzarella cheese
2 tblsp. fine dry whole wheat breadcrumbs
1 tsp. parsley flakes
½ tsp. dried whole oregano
½ tsp. dried whole basil
½ tsp. dried whole thyme
1 15-oz. can tomato sauce

Drain mushrooms, reserving juice; set mushrooms aside. Add enough water to juice to equal ½ cup. Combine mushroom liquid, ground beef, ½ cup breadcrumbs, egg, salt, and garlic powder; mix well. Cover and chill 30 minutes. Shape mixture into a 12 x 8-inch rectangle on a sheet of waxed paper. Sprinkle with mushrooms, bacon, onion, cheese, 2 tblsp. breadcrumbs, and seasonings, leaving a 1-inch margin around edges. Beginning at short end, roll meat up jellyroll fashion, lifting waxed paper to help in rolling. Press edges and ends together to seal. Place tomato sauce over top. Cover; bake at 350° for 1 hour. Let roll stand 5 minutes before slicing. Makes 6 servings.

HAMBURGER PIE

1 lb. lean ground beef
1 c. cooked cracked wheat
1⅓ c. mashed potato mix
1 c. milk
¼ c. catsup
1 egg, beaten
1 tblsp. instant minced onion
1 tsp. salt
⅛ tsp. pepper
1⅓ c. mashed ptotato mix
½ c. (2 oz.) shredded sharp Cheddar cheese

Combine first 9 ingredients in a large bowl; press mixture into a lightly greased 9-inch pie pan. Bake, uncovered, at 350° for 35 minutes. Prepare 1⅓ cups mashed potato mix according to package directions. Spoon on top of meat mixture; sprinkle with cheese. Bake an additional 4 minutes or until cheese melts. Makes 6 servings.

KANSAS BEEF SUPPER

2 tblsp. shortening
2 lbs. beef stew meat cut into 1-inch cubes
2 medium onions, sliced
1 c. water
2 large potatoes, pared and sliced thin
1 c. cooked wheat kernels
1 can (10½-oz.) cream of mushroom soup
1 c. dairy sour cream
1¼ c. milk
1 tsp. salt
¼ tsp. pepper
1 c. wheat cereal flakes, crushed

Melt shortening in large skillet. Add beef and onions. Cook and stir until meat is brown and onion is tender. Add water. Heat to boiling. Reduce heat, cover and simmer 45 minutes. Heat oven to 350°. Pour meat mixture into ungreased baking pan 13½ x 9 x 2-inches. Arrange potatoes and wheat kernels on meat. Stir together soup, sour cream, milk, salt and pepper. Pour over potatoes. Sprinkle with cheese and cereal. Bake uncovered 1½ hours. Makes 8 servings.

MEAT LOAF MOZZARELLA

2 lb. lean ground beef
2 eggs
1 c. whole wheat dry bread crumbs
1 can (8 oz.) tomato sauce
1 c. shopped onion, sauteed
⅓ c. grated Parmesan cheese
¼ c. green olives, chopped
1 tsp. salt
1 tsp. oregano
½ tsp. crushed red pepper
10 slices (2 x 4-inch) mozzarella cheese
5 rings of bell pepper, cut in half

Mix together ground beef, eggs, bread crumbs, tomato suace, onion, Parmesan cheese, olives, salt, oregano and red pepper. Pack mixture into 12-cup microwave Bundt pan. Rotating pan once every 4 minutes, microwave on 70 percent (medium-high) 18 to 20 minutes. Pour off any grease that may have accumulated. Let meat loaf stand 5 minutes before turning out onto microwave safe plate. Arrange mozzarella cheese over the top of the meat loaf and decorate with bell pepper. Sprinkle with additional crushed red pepper if desired. Microwave on 70 percent for 1 to 2 minutes, or until cheese begins to melt. Makes 6 to 8 servings. Freezes well.

QUICHE

3 eggs, slightly beaten
2 c. warm low-fat milk
½ tsp. salt
Pinch pepper
¾ c. grated Swiss cheese
1 whole wheat piecrust (8-inch) partially baked
Pinch nutmeg
1 tsp. butter

Preheat oven to 350°. Combine eggs, milk, salt, pepper. Scatter cheese evenly in bottom of pie shell and pour milk mixture over it. Sprinkle top with nutmeg and dot with butter. Place pie plate on cookie sheet and bake for ½ hour or until set. Remove when outside is set and middle couple of inches still jiggles if you tap pan. Quiche must stand for 10 to 15 minutes before cut; in that time, center will set. Try not to overbake, as texture will be less smooth. Serves 6.

SAUSAGE CASSEROLE

3 medium potatoes, cooked
¼ c. chopped onion
¼ c. margarine
¼ c. plus 2 tblsp. flour
½ tsp. salt
¼ tsp. pepper
¼ tsp. thyme, crushed
2 c. milk
½ lb. smoked sausage links, thinly sliced
1 c. frozen peas, rinsed and drained
1 c. cooked wheat kernels
3 tblsp. dry whole wheat breadcrumbs
1 tblsp. margarine, melted

Pare and slice enough potatoes to equal 2 cups; set aside. Cook onion in ¼ cup margarine until tender. Blend in flour and seasonings. Gradually add milk stirring until smooth. Cook, stirring constantly, until mixture is slightly thickened. Place half of sausage in bottom of 2-quart ovenproof casserole. Add half of potatoes, half of peas, half of wheat kernels and half of sauce. Repeat layers with remaining ingredients, ending with sauce. Sprinkle breadcrumbs mixed with melted margarine over top. Bake at 350° for 20 to 30 minutes, or until crumbs are lightly browned. Makes 4 servings.

STEAK STRIP BONANZA

1 lb. beef top round steak, cut 3/4 inch thick
1 can (14-oz.) tomatoes
1 c. thinly sliced carrots
1/2 c. chopped onion
1/2 tsp. basil leaves
1/4 tsp. oregano leaves
2 tblsp. stone ground whole wheat flour
1 tsp. salt
1/8 tsp. cumin
1/8 tsp. pepper
1 tblsp. oil
1 c. thinly sliced zucchini

Slice partially frozen top round steak in strips 1/8 inch or less thick and 2 to 2 1/2-inches long. Set aside. Drain tomato liquid into 4-cup glass measure. Add carrots, onion, basil and oregano to liquid; cover and microwave on high 4 minutes, stirring midway through cooking. Combine flour, salt, cumin and pepper; dredge steak strips in mixture. Add oil to 11 3/4 x 7 1/2-inch or 8 x 8-inch microwave safe baking dish; spread steak strips in layer over bottom. Cover and cook on 50 percent (medium) 6 minutes, stirring midway through cooking. Break up tomatoes and add with carrot mixture and zucchini to meat mixture. Cover and microwave on 50 percent 12 minutes, stirring every 3 minutes. Makes 4 servings. Serve over bed of cooked wheat.

SPICY BEEF AND WHEAT CASSEROLE

1 lb. ground beef
1 small onion, chopped
1 medium-sized green pepper, chopped
1 16-oz. can tomatoes, drained & chopped
1 13-oz. can french onion soup, undiluted
1 10 3/4-oz. can cream of mushroom soup, undiluted
1 c. uncooked cracked wheat
1 tsp. chili powder
Dash of hot sauce
1 c. (4 oz.) shredded Cheddar cheese

Combine beef, onion, and green pepper in a large skillet; cook until meat has browned, stirring to crumble meat. Drain off drippings. Add remaining ingredients except cheese; mix well. Spoon into a lightly greased 13 x 9 x 2-inch baking dish; cover, and bake at 350° for 1 hour. Remove from oven, top with cheese, and bake an additional 3 minutes or until cheese melts. Makes 6 servings.

STIR-FRIED SAUSAGE AND SQUASH

1 pkg. (8 oz.) brown and serve sausages, cut into quarters
1/4 lb. mushrooms, sliced
2 carrots, cut into 1-inch sticks (about 1 cup)
1 small zucchini, sliced (about 1 cup)
1 small yellow squash, sliced (about 1 cup)
1 c. cooked wheat kernels
1/3 c. sliced water chestnuts
1 c. chicken broth
1/4 c. dark corn syrup
2 tblsp. soy sauce
1 tblsp. vinegar
3 tblsp. cornstarch
1/4 c. water

In large skillet over medium heat, cook sausages 5 minutes or until browned. Remove from skillet. Add mushrooms, carrots, zucchini, yellow squash, cooked wheat, and water chestnuts. Stirring constantly, cook 1 minute. In small bowl stir together chicken broth, corn syrup, soy sauce and vinegar. Stir into vegetable mixture. Cover; cook 5 minutes or until tender-crisp. Add sausages. In small bowl stir together cornstarch and water. Stir into sausage mixture. Stirring constantly, bring to a boil and boil 1 minute. Serve over rice. Makes 4 servings.

TASTY TACO LOAF

2 lbs. ground chuck
½ c. dry whole wheat breadcrumbs
2 eggs, slightly beaten
1 medium onion, finely chopped
1 pkg. taco seasoning mix
1 can (16 oz.) refried beans
1 c. Cheddar cheese, grated
Additional Cheddar cheese for topping
Sliced avocado
Tomato slices
Black olives
Crushed Corn Chips

Preheat over to 350°. Mix ground beef, breadcrumbs, eggs, onion and taco seasoning mix. Mix well. On waxed paper, pat the mixture into a large hamburger patty. Spread with one-half of refried beans and then sprinkle grated cheese on top of beans. Roll into a loaf, place in a shallow baking dish, bake 1 hour and 15 minutes or until done. Heat the other half of the beans during the last 15 minutes of baking time and pour over the taco loaf after removing from the oven. Top with more grated cheese. Fix peeled, sliced avocado, tomato slices and black olives and arrange in a pretty design on top. Serve on a bed of crushed corn chips. Serves 6.

TACO MEAT BALLS

1 lb. lean ground beef
1 c. **each** finely chopped onions, green pepper and celery
2 c. cooked cracked wheat
2 eggs, beaten
2 tsp. garlic salt
1 8-oz. can taco sauce
11-oz. can condensed cheddar cheese soup

Combine meat, onions, green pepper, celery, wheat, eggs and garlic salt. Mix well and form into 12 meatballs. Place in lightly greased 2½ quart casserole. Bake at 350° for 30 minutes. Blend and heat taco sauce and soup. Pour over meat balls; cover and continue baking 30 minutes longer. Makes 6 servings. Make the meatballs tiny, and serve this recipe as a party appetizer, too.

WHEAT MEAT LOAF

1 c. cooked cracked wheat OR whole wheat kernels
1 lb. lean ground beef
½ lb. sausage
1 c. tomato juice or sauce
2 eggs
1 tsp. salt
½ tsp. pepper
1 c. whole wheat bread crumbs

Glaze:
½ c. catsup
3 tblsp. molasses
1 tsp. prepared mustard

Combine loaf ingredients well. Turn into an 8 or 9-inch square pan, shape and spread glaze over top. Bake in slow oven (325°) for 1 hour. Makes 8 to 10 servings.

WHEAT QUICHE

9-inch single whole wheat pie crust
½ c. sliced fresh mushrooms
¼ c. chopped green pepper
¼ c. shredded carrot
1 tblsp. butter OR margarine
4 beaten eggs
1 c. light cream
½ c. milk
½ tsp. salt
½ tsp. dried basil
½ c. shredded Monterey Jack cheese
1 c. cooked wheat kernels

Pre-cook pie crust at 450° for 5 minutes. Saute vegetables in butter just till tender. In a mixing bowl, stir together the beaten eggs, cream, milk, salt and dried basil. Stir in the cooked vegetables and cheese. Sprinkle cooked wheat kernels in bottom of pastry shell. Pour in egg mixture. Bake at 375° for 30 minutes or until knife inserted comes out clean. Cool on wire rack 5 minutes before serving. Makes 6 servings.

WHOLE WHEAT MAIN DISH

1 tblsp. butter
1 c. celery, chopped
¼ c. green pepper, chopped
½ c. onion, chopped
¾ c. roasted peanuts
2 c. cooked wheat
2 tblsp. soy sauce
1 tsp. salt

Melt butter, add celery, green pepper and onion. Saute until vegetables are clear. Add wheat and roasted peanuts then soy sauce and salt. Heat and serve.

ZUCCHINI ONE-DISH SUPPER

1 lb. ground beef
2 medium onions, chopped
1 green onion with top, chopped
1 10-oz. can tomatoes with green chiles
1 8-oz. can tomato sauce
1 6-oz. can tomato paste
1 medium-sized green pepper, chopped
4 medium zucchini, thinly sliced
½ c. cooked cracked wheat
½ tsp. garlic salt
½ tsp. dried whole oregano
¼ tsp. pepper
1 c. (4 oz.) shredded Cheddar cheese
½ c. grated Parmesan cheese

Cook ground beef and onions in a large skillet over medium heat until meat is browned, stirring to crumble meat. Drain well. Add next 4 ingredients; simmer 10 minutes, stirring occasionally. Add zucchini, cracked wheat, garlic salt, oregano, and pepper, and simmer 10 minutes. Stir in Cheddar cheese, and spoon into a lightly greased 13 x 9 x 2-inch baking dish. Sprinkle with Parmesan cheese. Bake at 350° for 20 minutes. Makes 6 servings.

Blessed is the person who is too busy to worry in the daytime and too sleepy to worry at night.

IDE DISHES & VEGETABLES

ALMOND VEGETABLES MANDARIN

1 c. carrots, thinly sliced
1 c. green beans, cut about 1 inch
2 tblsp. vegetable oil
1 c. cauliflower thinly sliced
½ c. green onions, sliced
½ c. cooked wheat kernels
1 c. water
2 tsp. chicken-style seasoning
2 tblsp. cornstarch
Pinch of garlic powder
½ c. whole almonds, unblanched

Cook and stir carrots and beans with oil in skillet over medium-high heat 2 minutes. Add cauliflower and onion; cook 1 minute longer. Add mixture of water, chicken seasoning, cornstarch, and garlic. Cook and stir until thickened. Vegetables should be crisp-tender. If they need further cooking, reduce heat, cover and steam to desired doneness. Add almonds and wheat. Recipe may be doubled only. Do not make larger quantity at one time. Serve in an attractive dish.

BAKED WHEAT AU GRATIN

1 tblsp. butter
3 tblsp. chopped onion
1 tblsp. flour
½ tsp. salt
½ tsp. dry mustard
1 c. milk
1 c. grated cheddar OR American cheese
2 c. cooked wheat kernels
Paprika

Melt butter; saute onion until yellow. Remove from heat. Blend in flour, salt and mustard, stir in milk. Return to heat and cook until thickened, stirring constantly. Add cheese, stir until melted. Stir in wheat. Pour into casserole and sprinkle with paprika. Cover and bake at 350° for 30 minutes until lightly browned. Makes 4 to 6 servings.

CRACKED WHEAT BAKED WITH RAISINS AND PINE NUTS

2 c. boiling water
1 c. cracked wheat
1 tsp. chopped fresh basil leaves OR ½ tsp. leaf basil, crumbled
1 tsp. salt
½ c. golden raisins
½ c. dark raisins
½ c. toasted pine nuts OR slivered almonds

Preheat oven to slow (300°). Butter 1-quart casserole that has a cover. Combine boiling water, cracked wheat, basil, salt and raisins in casserole. Cover top with aluminum foil. Place lid on top. Bake for 40 to 45 minutes or until liquid is absorbed and cracked wheat is cooked. Sprinkle with nuts.

CRACKED WHEAT-MUSHROOM PILAF

2 c. diced mushrooms
1 tblsp. fresh, chopped parsley
1 small onion, diced
2 c. cracked wheat
½ c. oil
4 c. chicken broth
Salt and pepper
1 bay leaf

Saute mushrooms, parsley, onion, and wheat in oil. Put into a 2-quart casserole with broth and seasonings, and cover. Bake at 350° for 45 minutes, or until cracked wheat is tender and liquid is absorbed. Serves 4 to 6.

CRACKED WHEAT WITH GREEN PEAS (Microwave Recipe)

2¼ c. hot water (165°)
1 c. uncooked cracked wheat
½ c. onion, finely chopped
1 clove garlic, minced
2 tblsp. butter OR margarine
2 tsp. instant chicken bouillon (optional)
½ tsp. salt
1 pkg. (10 oz.) frozen green peas

Mix all ingredients except peas in 2-quart casserole. Cover tightly; microwave on high (100%) 12 minutes. Stir in peas. Cover tightly and microwave on high (100%) until cracked wheat is tender and water is absorbed, 7 to 11 minutes longer. Use as a side dish like rice or noodles.

CHEESY BROCCOLI BAKE

10 oz. fresh OR frozen broccoli, chopped (4 cups)
1 can (10¾ oz.) cheddar cheese soup
1 can (6½ oz.) chunk ham, turkey OR chicken
1 c. cooked wheat kernels
½ c. sour cream
½ c. buttered bread crumbs

Preheat oven to 350°. Cook broccoli until tender, drain well. Stir in soup and sour cream, wheat and meat. Mix well and place in a 1½ quart casserole. Sprinkle with bread crumbs. Bake 40 minutes. Makes 6 servings.

GOLDEN ZUCCHINI BAKE

½ c. chopped onion
1 tblsp. butter OR margarine
5½ c. coarsely chopped zucchini
3 eggs, slightly beaten
½ c. milk
1 c. cooked wheat kernels
¾ tsp. oregano leaves, crushed
¼ tsp. ground white pepper
3 c. (bite-size) crispy wheat cereal crushed to 1 cup
1½ c. (6 oz.) shredded sharp Cheddar cheese, divided

Preheat oven to 325°. Butter shallow 1½-quart baking dish. In large skillet saute onion in butter about 5 minutes or until tender. Add zucchini. Saute 3 minutes longer. Drain if necessary. Set aside. In large bowl combine eggs, milk and seasonings. Stir in cereal crumbs, 1 cup cheese and zucchini mixture. Turn into baking dish. Sprinkle with remaining ½ cup cheese. Bake 35 to 40 minutes or until set and bubbly. Makes 6 servings.

KANSAS BAKED WHEAT

2 c. cracked wheat
⅓ c. ketchup
3 tblsp. finely chopped onion
2 tblsp. packed brown sugar
¼ tsp. dry mustard
2 slices uncooked bacon, diced

In 2-quart saucepan over high heat, bring 1 c. water* to a boil. Add cracked wheat, return to a boil. Reduce heat to medium-low. Cover and cook 5 minutes, or until wheat is tender. Remove from heat; drain well. Stir ketchup, onion, brown sugar and dry mustard into wheat in saucepan. Pour into greased 1½-quart casserole. Sprinkle with diced, uncooked bacon. Bake in 350° oven for 40 minutes, or until bubbly around edge and bacon is cooked. Makes 6 to 8 servings.
*Add more water if it gets dry before the wheat is done.

KING COLE CURRY

1 bunch broccoli (1½ lbs.)
½ head cauliflower (½ lb.)
1 medium potato
½ onion, chopped
2 tblsp. oil
2 tblsp. stone ground whole-wheat flour
Water
1 lb. tofu, rinsed and mashed
2 tblsp. lemon juice
1 tblsp. natural soy sauce
¾ tsp. curry powder
½ tsp. salt

Topping:
½ c. whole-wheat bread crumbs
2 tsp. oil
Pinch each pepper, paprika

Cut broccoli and cauliflower into flowerets, peel stems of broccoli and slice about ¼ inch thick. Cut potato in half lengthwise and slice ¼ inch thick. Steam vegetables together until just tender. Saute onion in 2 tblsp. oil. Add flour; cook 2 minutes over medium heat. Stir in 1 cup water and cook until thickened. Blend tofu, lemon juice, soy sauce, ½ c. water, curry powder and salt. Add to onion mixture. Add more salt or lemon juice to taste. Preheat oven to 350°. In casserole dish, spread half of vegetables, then half of sauce. Repeat for second layer. Combine bread crumbs with 2 tsp. of oil, pepper and paprika, and sprinkle over casserole. Bake covered 10 minutes, then uncovered 10 to 15 minutes more or until sauce is bubbly. Serves 6 to 8.

MACARONI & CHEESE

3 tblsp. oil
2½ c. uncooked whole wheat macaroni
½ tsp. salt
¼ tsp. pepper
½ lb. colby cheese
1 qt. milk

Put oil in casserole dish. Add macaroni and stir until coated with oil. Add diced cheese, salt, pepper and milk. Bake uncovered at 325° for 1½ hours.

OATMEAL / WHOLE-WHEAT STOVE-TOP STUFFING

1 c. uncooked old-fashioned OR quick-cook oats
10 slices whole-wheat bread, cubed
4 tblsp. butter OR margarine
3 celery stalks, sliced
2 eggs, slightly beaten
2 tblsp. butter OR margarine
1 medium onion, chopped
1 tblsp. parsley, chopped
¾ tsp. salt
¼ c. milk (more, if needed)
Poultry seasoning, sage, pepper, thyme, coriander or savory, season to taste

Melt 2 tablespoons butter or margarine in skillet over medium heat. Add oats and cook, stirring frequently, until lightly browned. Spoon oats into bowl and set aside. Melt 4 tablespoons butter or margarine in same skillet over medium heat. Add onion and celery and saute. Put oats in skillet, add bread cubes, mix eggs and milk and pour over cubes. Add remaining ingredients; gently toss to mix well. Heat through and serve. Serves 10.

Do be careful. Remember your tongue is in a wet place and is apt to slip.

PUFFY CHEESE FONDUE

Butter
1/4 c. butter
1/4 c. all-purpose flour
1 tblsp. instant minced onion
1 tsp. prepared mustard
1/2 tsp. salt
1/2 tsp. Worcestershire sauce
1 c. milk
1/2 c. (2 oz.) shredded cheddar cheese
1/2 c. (2 oz.) shredded swiss cheese
4 eggs, separated
3 slices whole OR cracked wheat bread, cut in 1/2-inch cubes
1/4 tsp. cream of tartar

Butter bottom and sides of 1 1/2-quart souffle dish. Make 3-inch band of triple-thickness aluminum foil long enough to go around dish and overlap 2 inches. Lightly butter 1 side of band. Wrap around outside of dish with buttered side in and fasten with string (collar should extend 2 inches above rim of dish). In large saucepan, melt 1/4 c. butter. Blend in flour, onion, mustard, salt and Worcestershire sauce. Cook, stirring constantly, over medium-high heat until mixture is smooth and bubbly. Stir in milk all at once. Add cheeses. Continue to cook and stir until mixture thickens and cheeses melt. Remove from heat. Let stand 5 minutes. Beat in egg yolks. In large mixing bowl, beat egg whites and cream of tartar at high speed until stiff but not dry, just until whites no longer slip when bowl is tilted. Gently but thoroughly fold yolk mixture and bread cubes into whites. Carefully pour mixture into prepared dish. Bake in preheated 350° oven until puffy, delicately browned and fondue shakes slightly when oven rack is gently moved back and forth, 35 to 40 minutes. Serve immediately. Makes 4 servings.

SWEET AND SOUR RED CABBAGE

1 onion
2 tblsp. oil
1 apple
8 c. (1 small head) shredded red cabbage
3/4 c. water OR vegetable stock
2 tblsp. whole wheat flour
2 tblsp. cider vinegar
1 tblsp. brown sugar
1 tsp. salt
1 tsp. caraway seeds

Chop the onion coarsely and saute in oil in a heavy pan until soft. Grate apple. Add to onion with cabbage and 1/4 c. water. Cook 5 minutes. In a jar shake together the flour, vinegar, sugar, salt and remaining 1/2 c. water. Add caraway seeds and stir into cabbage. Cook another 5 to 10 minutes, until cabbage is just tender. Makes 6 servings.

WHEAT / CORN CASSEROLE

About 8 ears fresh-picked corn
2 tblsp. regular margarine OR butter
1 medium green pepper, seeded and chopped
2 c. cooked wheat kernels
2 c. shredded Cheddar cheese
½ c. milk
1 (4 oz.) jar sliced pimentos, drained
½ tsp. salt
⅛ tsp. pepper

With sharp knife, cut corn kernels from cobs at two-thirds of their depth, then scrape cobs with dull side of knife to remove remaining pulp. Measure 4 c. corn with pulp; set aside. In 10-inch skillet over medium-high heat, melt margarine. Add green pepper; cook 2 minutes. Stir in corn. Cover and cook, stirring occasionally, 5 minutes, or until corn is tender-crisp. Pour into large bowl. Gently stir in remaining ingredients. Spread in greased 12x8x2baking dish. Bake in 325° oven 35 minutes, or until casserole is bubbly around edges and hot in center. Makes 12 servings.

WHEAT JARDIN

¾ c. chopped onions
1½ lbs. zucchini, thinly sliced
3 tblsp. butter
1 lb. can whole kernel corn, drained
1 lb. can tomatoes
3 c. cooked wheat kernels
1½ tsp. salt
¼ tsp. pepper
¼ tsp. ground coriander
¼ tsp. leaf oregano

Saute onions and zucchini in butter until tender; add remaining ingredients. Cover and simmer 15 minutes. Makes 8 servings.

WHEAT-NUT PILAF

1 c. pre-soaked wheat kernels OR cracked wheat
3 tblsp. butter OR margarine
2 c. beef OR chicken broth OR bouillon
2 medium carrots, shredded
½ tsp. salt
½ c. chopped walnuts, pecans OR almonds

In large ovenproof skillet or flameproof casserole, saute kernels in butter about 5 minutes to brown lightly, stirring occasionally. Stir in broth, carrots, and salt; bring to boil. Cover and bake in 350° oven for 25 minutes or until broth is absorbed, stirring occasionally. Stir in choice of nuts. Makes 4 servings.

WHEAT AND NUT CASSEROLE

- 2 c. cooked wheat kernels
- 2 c. walnuts OR pecans, chopped
- 1/4 c. green peppers, chopped
- 1 small onion, chopped OR
- 2 tblsp. dry onion, minced
- 1/2 c. celery, chopped
- 1 c. dry, seasoned whole-wheat bread crumbs
- 3 bouillon cubes dissolved in a little water OR Kitchen Bouquet
- 2 tblsp. soy sauce
- 2 tblsp. vegetable oil
- 2 tblsp. whole-wheat flour
- 3/4 to 1 1/2 c. milk (as needed to give desired consistency)

Mix all ingredients well. Place in oiled casserole baking dish or individual molds. Bake at 350° one hour. Unmold immediately. Garnish with carrot curls, parsley, tomato wedges, etc.

WHEAT OVEN PILAF

- 3 c. water
- 1 c. cracked wheat
- 1/2 tsp. salt (optional)
- 1/2 c. finely chopped onion
- 1/2 c. shredded carrot
- 1/2 c. chopped green onions
- 2 tblsp. cooking oil
- 1/4 c. toasted wheat germ
- 1/4 tsp. garlic powder
- 2 tblsp. snipped parsley

In a medium saucepan bring water to boiling. Add wheat and salt; return to boiling. Reduce heat; cover and simmer for 12 to 15 minutes or till wheat is tender. Do not drain. Meanwhile, in another medium saucepan cook onion, carrot, and green onions in cooking oil till tender. In a greased 1 1/2-quart casserole combine undrained wheat, vegetables, wheat germ and garlic powder. Bake, uncovered, in a 350° oven for 25 to 30 minutes or till light brown. Fluff with a fork to serve. Sprinkle with parsley.

WHEAT RING AROUND

- 4 tblsp. margarine
- 6 c. cooked wheat kernels
- 1/2 c. evaporated milk
- 2 large eggs, beaten
- 1 c. chopped mushrooms, cooked
- 1/4 c. chopped green onions with tops

Put margarine into 12-cup microwave Bundt pan and microwave on high 1 minute or until melted. Tilt pan around to coat all sides. Pour excess margarine into wheat. Mix milk, eggs, chopped mushrooms and green onions with wheat. Pack into prepared pan. Cover with wax paper and microwave on 70 percent (medium-high) 11 to 12 minutes, rotating dish midway through cooking. Let stand 5 minutes before inverting onto serving plate. Serves 12.

SPINACH-WHEAT CUSTARD

2 10-oz. pkgs. frozen, chopped spinach
6 eggs, slightly beaten
3 tblsp. melted butter
½ c. grated onion
1 tsp. salt
¼ tsp. pepper
1 tblsp. vinegar
½ c. grated Parmesan cheese
1½ c. scalded milk
2 c. cooked wheat kernels
Sauteed onion rings for garnish

Cook spinach following directions on package. Drain well. Combine eggs, butter, onion, salt, pepper, vinegar and cheese. Stir in milk. Fold in spinach and wheat. Pour mixture into buttered 2-quart baking dish. Set dish in pan of hot water, making sure water comes almost to top of casserole. Bake at 350° for 1 hour, or until custard is set. Center may still be slightly soft, but custard will continue cooking after casserole is removed from oven. Top custard with onion rings for garnish. Delicious served as a side dish with a roast.

SALADS

BROWN BERRY & APPLE SALAD

3 c. cooked wheat kernels
1 c. chopped celery
Caraway dressing (recipe follows)
2 small red-skinned apples, cored & diced
1/4 lb. Swiss cheese, julienne strips
1/4 c. finely chopped parsley

In a salad bowl, combine cooked wheat, celery and dressing. Cover and chill at least one hour or overnight. Stir in apples. To serve, garnish with Swiss cheese strips and parsley. Serves 6.

CARAWAY DRESSING

In a small bowl, mix together 1/2 c. mayonnaise, two tablespoons each coarse ground prepared mustard and white wine vinegar and one teaspoon caraway seed.

CARROT & PINEAPPLE SALAD

2 1/2 c. carrots, shredded
1 can (8 oz.) crushed pineapple, drained
2 c. cooked whole wheat kernels
1/2 c. salad dressing OR mayonnaise

Combine ingredients and chill.

CHEESE APPLE SALAD

3/4 c. dairy sour cream
1 tsp. fresh lemon juice
1/4 tsp. salt
2 medium tart apples
1/2 c. diced Swiss cheese
1/2 c. shredded Cheddar cheese
1/2 c. chopped celery
1/3 c. halved red grapes
1/4 c. coarsely chopped walnuts
Salad greens
1/2 c. cooked wheat kernels

Blend lemon juice and salt into sour cream. Chill. Cut apples into thin slices leaving peel on; sprinkle with lemon juice. (For a more colorful salad use combination of red and green apple.) Add cooked wheat, Swiss and Cheddar cheeses, celery, grapes and walnuts; toss together. Just before serving, fold in sour cream to blend. Serve on salad greens, garnished with apple slices and grapes, if desired. Makes 4 cups.

CHEESY BAKED POTATO SALAD

8 strips bacon, cubed
1 c. chopped celery
1 c. chopped onion
3 tblsp. whole wheat flour
1½ c. water
1 c. cider vinegar
⅔ c. sugar
1 tsp. salt
¼ tsp. pepper
8 c. cubed cooked potatoes
1 jar (2 oz.) diced pimentoes (optional)
1 c. shredded sharp cheddar cheese

Preheat oven to 350°. Fry bacon in 10 inch skillet until crisp. Remove and drain on paper towels. Drain fat and measure ¼ c. of fat and return to skillet. Add celery and onion; cook 1 minute. Blend in flour and stir in water and vinegar. Cook, stirring constantly until mixture is thick and bubbly. Stir in sugar, salt and pepper; pour mixture over potatoes and bacon in greased 3 quart casserole. Toss the pimentoes lightly with the potatoes and bacon mixture. Cover and bake for 45 minutes. Sprinkle with cheese before serving.

CHEESY ORANGE SALAD

1 pkg. orange jello
1 c. boiling water
2 c. miniature marshmallows
1 (6 oz.) can frozen orange juice
½ c. sugar
1 (3 oz.) pkg. cream cheese
1 (8 oz.) can crushed pineapple
1 c. cottage cheese
1 c. cooked wheat kernels
2 bananas, mashed
1 can mandarin oranges
1 c. cream, whipped

Pour boiling water over jello, add marshmallows and stir until jello dissolves. Defrost orange juice and stir in sugar. Add to jello mixture. Cream the cheese. Add crushed pineapple to cheese and add to jello mixture, then add cottage cheese, wheat, bananas and mandarin oranges. Stir until well mixed. Chill until slightly set; fold in whipped cream. Put in refrigerator 4-5 hours before serving.

Have two goals: wisdom—that is knowing and doing right—and common sense. Don't let them slip away. Prov. 3:21

CHICKEN-FRUIT SALAD

3 c. chicken, cooked & chunked
¾ c. celery, chopped
¾ c. red grapes, halved & seeded
20 oz. can pineapple chunks, drained
11 oz. can mandarin oranges, drained
¾ c. cooked wheat kernels
2 T. nuts, chopped
¼ c. salad dressing
Lettuce Leaves

Toss chicken, celery, grapes, pineapple, oranges and wheat together lightly. Gently mix salad dressing with chicken mixture. Chill. Serve on lettuce leaves. Garnish with nuts. Makes six 1 cup servings.

CHICKEN SALAD ORIENTAL

1½ c. chopped, cooked chicken
2 16 oz. cans mixed Chinese vegetables, rinsed & drained
1 c. cooked wheat kernels
2 Tblsp. diced onion
1 c. chopped celery
½ c. chopped green pepper
1 tsp. salt
2 tsp. soy sauce
¾ c. mayonnaise
Slivered, blanched almonds
Lettuce Leaves

Combine first seven ingredients. Combine soy sauce and mayonnaise and stir into salad. Chill at least 1 hour. Line glass bowl with lettuce leaves and fill with salad. Top with almonds. Serve with fresh fruit and whole wheat muffins for a delicious luncheon.

CHICKEN SALAD SUPREME

4 chicken breasts, cooked
1 c. celery, diced
¼ c. onion, diced
1 can mandarin oranges, drained
2 apples, diced
1 c. cooked wheat kernels
1 tsp. curry powder
1 tblsp. mustard
1 c. mayonnaise
½ c. pickle relish
1 c. slivered almonds, toasted

Remove skin from chicken breasts and dice. Toss chicken, celery, onion, oranges and apple. Combine curry, mustard, and pickle relish with mayonnaise, stirring until well blended. Fold into chicken salad. Add a little cream or additional mayonnaise until salad is as moist as you desire. Salt to taste. Chill until serving. To serve, pile into a lettuce-lined glass serving bowl. Top with almonds. For individual servings, arrange on a leaf of lettuce and accompany with a section of cantaloupe when in season.

DISAPPEARING CARROT SALAD

⅔ c. dried unsweetened coconut
½ c. walnuts
4 medium-large carrots (1 quart) grated
2 apples, peeled, quartered, and grated
1 c. cooked wheat kernels
Finely grated peel and juice of 1 lemon
1 c. orange OR tangerine juice
1 c. dried currants
Dash salt
1 tblsp. (more to taste) grated gingerroot

Toast coconut and walnut pieces in 300° oven. The coconut will take 5 minutes, the walnuts 10. Chop walnuts. Combine grated carrots, apples, wheat, orange and lemon juice, currants, salt and ginger. Add walnuts and coconut and serve, chilling first if preferred. Makes 8 servings.

GREEN PEA WHEAT SALAD

1 c. cooked wheat kernels
1 pkg. (12 oz.) frozen OR 1½ c. fresh peas, cooked, drained, cooled
1 tblsp. minced green onion
¼ c. diced celery
½ c. diced cheddar cheese
Mayonnaise to moisten

Combine all ingredients. Refrigerate one hour before serving.

HOT CHICKEN SALAD

3 c. cooked wheat kernels
3 c. diced cooked chicken OR turkey
2 c. diced celery
¼ c. chopped dill pickles
⅓ c. chopped nut meats
¼ c. grated onion
1½ tsp. salt
¼ tsp. pepper
1 c. mayonnaise
1 c. grated Cheddar cheese
1 c. crushed rice cereal OR corn flakes

Combine all ingredients except cheese and cereal. Turn into a buttered shallow 2½ quart casserole. Mix cheese and cereal. Sprinkle over top of salad. Bake at 350° for 30 minutes. Makes 8 servings.

There is one thing to be said about ignorance. It gives rise to 90% of the conversations.

PEANUTTY WHEAT SALAD

1 large can chunk pineapple
2 c. miniature marshmallows
Juice from pineapple
½ c. sugar
1 tblsp. flour
1 egg, beaten
1½ tblsp. vinegar
1 large tub Cool Whip
2 large unpeeled apples, cubed
1 c. peanuts
1 c. cooked wheat kernels

Drain pineapple, add marshmallows. Let stand overnight in refrigerator. Mix pineapple juice, sugar, flour, egg, vinegar and cook until thick. Cool in refrigerator. Add Cool Whip. Mix with pineapple and marshmallows and add wheat, peanuts and apples.

PERSISTENCE VEGETABLE SALAD

Salad:
1 can small green peas
1 can French style green beans
1 can white corn
½ c. chopped onion
1 c. chopped celery
2 green peppers, chopped
1 oz. chopped pimentoes
1 c. cooked wheat kernels

Dressing:
1 scant cup sugar
½ c. vegetable oil
⅔ c. cider vinegar
Salt, to taste
Pepper, to taste

Heat the dressing, and after it has cooked, pour over the combined salad ingredients.

PIZZA SALAD

1 jar (6 oz.) marinated artichoke hearts
1 bunch romaine lettuce
1 head iceberg lettuce
¼ c. vinegar
1½ tsp. crushed oregano
½ tsp. salt
1 clove garlic, crushed
1 green pepper, cut in strips
1 pkg. (10 oz.) Mozzarella cheese, cut in julienne strips
6 slices salami, cut in eights
12 cherry tomatoes, halved
¼ pound thinly sliced fresh mushrooms
½ c. cooked wheat kernels

Drain artichoke hearts, reserving marinade. Remove outer leaves of romaine to line salad bowl. Tear 3 cups romaine and 5 cups lettuce; toss together and chill. To prepare dressing: Combine marinade, vinegar, oregano, salt and garlic. Set aside to blend flavors. To serve: Line a large salad bowl with romaine leaves. Fill with chilled greens. Arrange artichoke hearts, green pepper, cheese, salami, tomatoes, mushrooms and wheat over the top. Pour dressing over salad. Makes 8 servings.

SCRUMPTIOUS CHICKEN SALAD

1½ c. uncooked macaroni rings
1½ c. cut-up cooked chicken
⅔ c. finely chopped celery
1 can (17 oz.) small early peas, drained
1 can (13¾ oz.) pineapple tidbits, drained
½ c. cooked wheat kernels
¼ c. slivered almonds
2 tblsp. diced pimento
1 to 1¼ c. mayonnaise OR salad dressing
2 tblsp. lemon juice
1¼ tsp. seasoned salt
¼ tsp. pepper

Cook macaroni rings according to package directions; rinse and drain. In a large bowl, combine macaroni, chicken, celery, peas, pineapple, wheat, almonds and pimento. Combine remaining ingredients in small bowl; gently stir into chicken mixture. Cover and chill several hours. If desired, garnish with paprika, fresh parsley or almonds at serving time. Makes 6 servings.

SO GOOD WHEAT SALAD

2 c. cooked wheat kernels
1 can (6 oz.) tuna OR chicken drained
½ c. celery, thinly sliced
½ c. chopped sweet pickles
½ tsp. salt
2 chopped boiled eggs
⅛ tsp. black pepper
¼ c. mayonnaise OR salad dressing

Toss ingredients together. Chill 4 or more hours. Makes 4 to 6 servings.

SPROUT SALAD

2 c. sprouts (a mixture of different sprouts tastes best)
2 tomatoes, chopped
2 stalks celery, chopped
2 green onions, chopped
1 carrot, grated
1 cucumber, chopped
½ c. parsley

DRESSING

½ c. safflower oil
½ c. apple cider vinegar OR lemon juice
¼ t. garlic powder
¼ t. tamari soy sauce

Combine all vegetables in large bowl. Mix well. Combine dressing ingredients and pour over salad. Marinate before serving.

SUMMER'S SPECIAL WHEAT FRUIT SALAD

- 2 c. cooked wheat kernels
- 1/4 c. lemon juice
- 4 tblsp. honey
- 2 peaches OR nectarines, seeded and chopped
- 1 c. blueberries
- 1 c. sliced strawberries

Put wheat in bowl; cover and chill. Blend lemon juice and honey. Add to wheat and mix well. Chill. Before serving, stir in peaches, blueberries and strawberries. Makes 8 to 10 servings.

SUPER SALAD BOWL

- 1 c. water
- 1 c. cooked wheat kernels
- 2/3 c. plain yogurt
- 1/2 tsp. dried dillweed
- Several dashes bottled hot pepper sauce
- 1 3 3/4-oz. can salmon, drained, skin and bones removed, and flaked
- 1/4 c. chopped celery
- 2 tblsp. sliced almonds
- 2 small heads Bibb lettuce
- 1/2 c. finely shredded carrot
- 6 radishes, shredded
- Crinkle-cut carrot and celery sticks (optional)
- Sesame breadsticks (optional)
- Lemon peel flowers (optional)
- Carrot tops (optional)

For dressing: Combine the plain yogurt, dillweed, and the hot pepper sauce. In a medium mixing bowl stir together the wheat kernels, flaked salmon, chopped celery, and sliced almonds. Divide the yogurt mixture in half. Add half of the dressing to the wheat mixture; toss gently till just mixed. Cover; chill. Chill the remaining yogurt dressing separately till serving time. To serve, remove centers from the small heads of lettuce, leaving outer leaves intact to form a bowl. Reserve centers of the lettuce for another use. On individual serving plates place the hollowed lettuce bowls. Arrange the shredded carrot, shredded radishes, and the chilled wheat mixture in the lettuce bowls. Arrange the carrot sticks, celery sticks and breadsticks around each filled lettuce bowl. Garnish with lemon peel flowers and carrot tops. Serve with the remaining dressing. Makes 2 main-dish servings.

TABOULI SALAD

1 lb. cracked wheat
4 lg. fresh tomatoes, peeled
2 medium cucumbers
3 OR 4 bunches green onions, tops included
2 lg. green peppers
2 bunches parsley
Salt to taste
1 c. vegetable oil
1 c. lemon juice

Empty wheat into pan, rinse in cold water and drain well. Chop vegetables as fine as possible. Combine drained wheat and chopped vegetables in large bowl. Mix oil, lemon juice and salt together. Pour over wheat and vegetables, Mix well. Refrigerate several hours before serving. If mixture is too dry, add tomato juice as desired.

TABOULI, II

2 c. cracked wheat
1 red onion OR 1 bunch green onions
1 green pepper
2 bunches parsley
2 tomatoes
1 cucumber
1 tsp. salt
1 c. lemon juice
1 c. vegetable oil

Rinse and drain cracked wheat. Allow to soften one hour. Mix lemon juice with wheat. Chop onions and green pepper. Add to wheat along with the oil and salt. When ready to serve, add chopped parsley, chopped tomatoes, and chopped cucumber. Toss well.

TANGY BEAN SALAD

½ c. cider vinegar
¼ c. vegetable oil
3 tblsp. sugar
½ tsp. salt (optional)
½ c. chopped onion
½ c. chopped green pepper
1 15-oz. can dark red kidney beans, drained
1 16-oz. can cut green beans, drained
1 15-oz. can light red kidney beans, drained
Sliced red onion rings (optional)
1 c. cooked wheat kernels

Combine vinegar, oil, sugar and salt. Add onion, green pepper, cooked wheat and drained beans; toss gently. Cover, chill several hours or overnight. Garnish with onion rings, if desired. Serve using a slotted spoon. Makes 10 servings.

WALDORF WITH WHEAT

2 tblsp. mayonnaise OR salad dressing
½ c. plain yogurt
¼ tsp. vanilla
⅛ tsp. salt
1 c. cooked wheat kernels
½ cup thinly sliced celery
2 medium apples, cored and sliced
Broken walnut pieces (optional)

Stir together the dressing, yogurt, vanilla and salt. Add the celery and wheat and chill, covered, until ready to serve. Stir in apples and nuts just before serving. Makes 4 servings.

WATERCRESS AND HARD-COOKED EGG SALAD

2 bunches firm watercress
1 tblsp. Dijon-style mustard
2 tblsp. red-wine vinegar
Salt to taste if desired
Freshly ground pepper to taste
½ c. cooked wheat kernels
6 tblsp. corn, peanut OR vegetable oil
2 hard-cooked eggs, chopped
2 tblsp. coarsely chopped red onion
1 tblsp. finely chopped parsley

Trim off and discard tough ends of watercress. Rinse leaves and tender stems. Drain well and pat dry. Put mustard, vinegar, salt and pepper in salad bowl. Blend with wire whisk. Add oil, beating with whisk. Add watercress, wheat, eggs and onion. Sprinkle with parsley and toss. Makes 4 servings.

WHEAT SALAD

¾ c. wheat sprouts
½ c. cooked wheat kernels
3 tblsp. chopped sweet pickle
½ tsp. salt
⅛ tsp. pepper
1½ c. thinly sliced celery
½ c. chopped green pepper
⅔ c. mayonnaise
2 hard-cooked eggs, chopped
1-2 tblsp. minced onion

Combine sprouts, dressing, sweet pickle, salt and pepper; chill. Prepare and combine remaining ingredients; chill. Combine mixtures just before serving. Makes 4 to 6 servings.

WHOLE WHEAT SALAD

4 c. cooked wheat kernels
2 c. small marshmallows
1 can (20 oz.) crushed pineapple and juice
1 c. chopped pecans
1 envelope whipped topping mix OR 2 c. whipped topping

Mix all ingredients well. Serves 20. This healthy and nutritious salad may be used as a dessert.

PIES

APPLE PIE WAI-KI-KI

1 9-in. unbaked whole wheat pie crust
4 c. pared, sliced tart apples
¾ c sugar
1 tblsp. quick cooking tapioca
¼ tsp. salt

Topping:
¼ c. sugar
2 tblsp. cornstarch
8¾ oz. crushed pineapple and juice
2 tblsp. lemon juice
1½ c. dairy sour cream

Pile apples into crust. Mix sugar, tapioca and salt. Sprinkle over apples. Cover edge of pie shell with strip of foil to prevent excess browning. Remove foil during last 10 minutes of baking. Bake at 400° for 40 minutes, or until apples are tender. Combine sugar, cornstarch, pineapple and juice and lemon juice. Heat just to boiling, stirring constantly. Spread over baked pie. Cool. To serve, top with sour cream, or sweetened whipped cream if you desire.

APPLE DELIGHT (PIZZA PIE)

Crust:
1¼ c. stone ground whole wheat flour
1 tsp. salt
½ c. shortening
1 c. shredded cheese (not processed)
¼ c. ice water

Filling:

½ c. powdered non-dairy cream
½ c. brown sugar
½ c. sugar
⅓ c. flour
¼ tsp. salt
1 tsp. cinnamon
½ tsp. nutmeg
¼ c. butter
6 c. apples, sliced
2 tblsp. lemon juice

Blend flour, salt and shortening until crumbly, add cheese, Sprinkle water over flour mixture and stir until dough forms. Roll pastry to fit a 15 x 10 x 1-inch jelly roll pan, shaping dough into corners and up sides of pan. Combine powdered cream, sugars, flour, salt and spices. Sprinkle half of this mixture over unbaked crust. Cut butter into remaining mixture until crumbly. Set aside for topping. Arrange pared apple slices overlapping edges in rows across pan, covering entire crust. Sprinkle apples with lemon juice, then cover with crumbs set aside for topping. Bake at 450° for 30 minutes or until apples are tender. Serve warm, plain or with a dollop of whipped cream or ice cream.

APPLE PIE WITH WHOLE-WHEAT CRUST

Whole-wheat pastry (recipe follows)
¾ c. brown or white sugar
2 tsp. cinnamon
¼ tsp. nutmeg
¼ tsp. ground cardamon
8 c. tart apples, cored, peeled, and thinly sliced
2 tsp. lemon juice
1 tblsp. butter

Combine sugar and spices. Add apples, stirring until well coated. If apples are extra juicy, sprinkle on one or two tablespoons whole-wheat flour, tossing well. Sprinkle with lemon juice. Roll out half of pastry to about ⅛-inch thickness and fit into buttered 10-inch pie pan. Pile apple mixture into pastry shell, mounding apples higher in center. Dot dabs of butter over apples. Roll out remaining pastry. Place over apples. Fold edges under, seal, and flute. Cut steam vents. Bake in preheated 425° oven 15 minutes. Reduce heat to 375° and bake about 35 minutes longer or until crust is golden. Serve warm with slices of cheddar cheese. Eight to ten servings.

WHOLE WHEAT PASTRY

1¼ c. stone ground whole wheat flour
¾ c. all-purpose flour
¾ tsp. salt
¾ tsp. baking powder
⅓ c. vegetable shortening
⅓ c. unsalted butter
3 to 5 tblsp. ice water

Sift flours, salt, and baking powder together into large bowl. Cut in shortening and butter until dough is crumbly and the size of small peas. Drizzle ice water over mixture, one tablespoon at a time, tossing lightly with fork until dough is moist enough to hold together, but is not sticky. Shape dough into smooth ball. Wrap tightly in plastic wrap and refrigerate at least 30 minutes before rolling. Makes enough pastry for two-crust pie.

CHEESE CRUMBLE APPLE PIE

1 unbaked 9-inch whole wheat pie shell
Topping:
½ c. all-purpose flour
⅓ c. sugar
⅓ c. firmly packed brown sugar
½ tsp. cinnamon
5 tblsp. butter
Filling:
5-6 c. peeled, sliced, cooking apples
1 tblsp. fresh lemon juice
1½ c. (6 oz.) shredded Cheddar cheese
4 tsp. all-purpose flour
¼ tsp. nutmeg

Preheat oven to 375°. Make a high rim on pie crust. To prepare topping, combine flour with sugars and cinnamon; cut in butter. Set aside. To prepare filling, toss together apples and lemon juice; mix together cheese, flour, and nutmeg; toss with apples. Arrange apples in pie crust. Sprinkle on topping. Bake 40 to 50 minutes.

CHOCOLATE CRUMB CRUST

1½ c. fine dry whole wheat breadcrumbs
¼ c. sugar
½ c. melted butter
1 oz. baking chocolate, melted

Combine breadcrumbs, sugar, butter and chocolate. Spread in buttered 9-inch pie pan; press firmly in place. Fill with a cooked filling and refrigerate until firm or if a meringue is desired, put on top of hot filling and bake at 400° until lightly browned. Cool out of a draft.

CRUMB TOPPING

½ c. rolled oats
¼ c. brown sugar
2 tblsp. chopped nuts
⅓ c. stone ground whole wheat flour
2 tblsp. wheat germ
⅛ tsp. salt or salt substitute
2-3 tblsp. orange juice (or any other fruit juice)

Combine all dry ingredients. Mix in orange juice until all is moistened. Sprinkle over any fruit pie.

MILE HIGH FRESH STRAWBERRY PIE

2½ c. fresh strawberries
¼ c. sugar
2 tblsp. water
1 c. sugar
2 egg whites
1 tsp. lemon juice
⅛ tsp. salt
½ c. whipping cream
1 tsp. vanilla
1 whole wheat pie shell, baked

Blend the first three ingredients in a mixer, food processor, or blender. Then combine the strawberry mixture, 1 c. sugar, egg whites, lemon juice, and salt; whip with electric mixer for 15 minutes, until very stiff but not dry. Whip cream; add vanilla. Fold into strawberry mixture by hand, until just blended. Pile in baked pie shell and freeze for several hours, or overnight. Note: one 10 oz. pkg. of frozen strawberries may be substituted for the first three ingredients.

Today's trying times will become tomorrow's good old days.

NEAPOLITAN WHEAT PIE

1 c. wheat kernels, soaked overnight in warm water
1 c. milk
1 c. sugar
⅛ tsp. salt
2 slices fresh lemon rind
2 tblsp. orange peel
6 eggs - separated
1 lb. ricotta cheese
½ tsp. cinnamon

Cream Filling:
3 egg yolks
¼ c. flour OR cornstarch
2 strips lemon rind
½ c. sugar
1 c. milk
½ tsp. vanilla

Drain wheat, place in pan and cover with water. Simmer 20 minutes. Remove from heat and let stand. Prepare filling by mixing the egg yolks and sugar. Gradually add the flour and mix well. Slowly add the milk, stirring constantly until smooth. Add lemon peel and cook over low heat until thickened. Remove from heat, discard lemon peel, add vanilla, stir and set aside. Drain wheat, add milk and 1 tblsp. sugar and lemon rind. Bring to boil and simmer until liquid is absorbed. Let cool. Strain the ricotta or blend with a blender. Stir in the sugar. Add egg yolks, orange peel and cinnamon. Fold into cream filling. Beat egg whites until stiff and fold into mixture. Spoon into 10-inch pie shell. Top with a lattice pie crust. Bake for 1 hour at 350°. For full flavor cool overnight in refrigerator. Serve at room temperature.

OATMEAL PIE

1 9-inch unbaked whole wheat pie crust
3 eggs, well beaten
⅔ c. sugar
1 c. brown sugar
2 tblsp. butter OR margarine
⅔ c. quick oatmeal
⅔ c. coconut
1 tsp. vanilla
⅔ c. wheat germ
12 pecan halves (optional)

Beat eggs and add sugar, melted butter, and vanilla. Add oats, coconut, and wheat germ to egg mixture. Stir well. Put in unbaked pie crust. Arrange pecan halves in circle close to center of pie. Bake 35-40 minutes at 350°.

PEACH PIE

Pastry:
1¼ c. all-purpose flour
AND
¾ c. stone ground whole wheat flour
1 tsp. salt
¾ c. shortening
5-6 tblsp. ice water

Combine flours and salt in mixing bowl. Using a pastry blender, or two knives, cut shortening into flour until it resembles small peas. Add half of ice water at a time and chase around the bowl with a fork, then add other half. Dough should cling together to form a ball. Divide in half and form into flattened balls. Roll out one portion on floured surface to a circle 1½ inches larger than inverted 9-inch pie pan. Fit loosely into pan. Fill with prepared peaches. Roll out remaining dough. Cut in slits for escaping steam. Place top crust over filling and trim ½ inch beyond rim of pan. Seal edge by folding top crust under bottom crust. Flute edge. Brush top of pie with milk. Bake in preheated 450° oven for 10 minutes, then 350° for 30 minutes or until a fork is slipped in slit and fruit feels soft. Cool on rack before slicing. Makes one 9-inch pie.

Peach Filling:
4-5 c. sliced peaches
½ c. sugar
¼ c. brown sugar
3 tblsp. cornstarch OR 2 tblsp. tapioca
½ tsp. almond extract
¼ tsp. each cinnamon and nutmeg (optional)
2 tblsp. butter OR margarine

Stir together dry ingredients and mix lightly with sliced peaches. Turn into pastry lined shell. Dot with butter. Cover with top crust which has slits cut in it. Follow directions for pastry. Variations: 36-40 oz. frozen sliced peaches — partially thaw and drain, sugar to taste or use 1½ cup sugar; 2 c. canned sliced peaches, drained. Decrease sugar to ½ cup.

PUMPKIN PIE SQUARES

Crust:
½ c. all-purpose flour
½ c. stone ground whole wheat flour
½ c. quick cooking rolled oats
½ c. brown sugar, packed
½ c. butter OR margarine

Filling:
2 c. (1 lb. can) pumpkin
1 can (13 oz.) evaporated milk
2 eggs beaten
¾ c. sugar
1 tsp. cinnamon
½ tsp. salt
½ tsp. ginger
¼ tsp. cloves

Topping:
½ c. chopped pecans
½ c. brown sugar
2 tblsp. butter OR margarine

Preheat oven to 350°.
Crust: Combine flour, oats and brown sugar. Cut in butter. Press in ungreased 9 x 13-inch cake pan. Bake 15 minutes. Filling: Combine ingredients together well. Pour over crust and bake another 20 minutes. Topping: Combine until crumbly and sprinkle over pumpkin filling. Return to oven and bake 15 to 20 minutes or until filling is set. Cool and serve plain or with whipped topping.

SQUASH-CHEESE PIE

Whole wheat pie crust for one 9-inch pie
2 3-oz. pkg. cream cheese, softened
½ c. packed light brown sugar
1½ tsp. cinnamon
¾ tsp. ginger
½ tsp. salt
¼ tsp. nutmeg
1 c. milk
3 eggs
1 12-oz. pkg. frozen cooked squash, thawed
1 tblsp. grated orange peel

Prepare pie crust. Line pie plate with pastry; trim edge, leaving ½-inch overhang; fold overhang under. With back of floured fork, press pastry firmly to rim of pie plate. Preheat oven to 375°. In large bowl with mixer at low speed, beat cream cheese, brown sugar, cinnamon, ginger, salt and nutmeg until smooth. Add milk, eggs, and thawed squash and continue beating until well mixed, constantly scraping bowl with rubber spatula. Place pastry lined pie plate on oven rack; pour in squash mixture. Bake 45 minutes or until knife inserted 1 inch from edge comes out clean. Cool pie completely on wire rack. Cover and refrigerate. To serve, garnish top of pie with orange peel. Makes 10 servings.

EASY WHOLE WHEAT PIE CRUST

1¼ c. stone ground whole wheat flour
½ tsp. salt or salt substitute
⅓ c. vegetable oil
4 tblsp. ice water

In mixing bowl combine flour and salt. Combine the oil and water and add all at once to the flour mixture. Stir briefly until mixed and then gather together into a ball. Roll out between two sheets of waxed paper with rolling pin (or pat out by hand directly into pie plate). Remove top piece of waxed paper. Lift pie crust and bottom waxed paper together and invert over pie plate. Peel off waxed paper. Fit the pie crust down into the pie plate and trim edges. For recipes calling for prebaked pie crust, bake crust in oven (425°) for 10 minutes and prick well before placing in the oven.

LOWER-FAT PIE CRUST

1 c. stone ground whole wheat flour
½ tsp. salt
3 tblsp. oil
¼ c. ice water

Stir dry ingredients together. Mix in oil and enough of water to make dough form ball. Roll flat between sheets of waxed paper and lift into pan. Make decorative edge. Bake at 400° for 10 minutes or until slightly browned and crisp. Makes one 9-inch shell.

WHOLE WHEAT CRUMB TOPPING

½ c. stone ground whole wheat flour
⅓ c. brown sugar
3 tblsp. shortening

Combine flour and sugar. Cut in shortening with pastry blender. Sprinkle over any fruit filling and bake as recipes indicate.

WHOLE WHEAT PASTRY

¾ c. vegetable shortening OR ⅔ c. lard
1 c. stone ground whole wheat flour
1¼ c. all-purpose flour
1 tsp. salt
4-5 tblsp. ice water

Cut shortening into flour and salt with pastry blender or 2 knives until particles are size of small peas. Sprinkle in ice water a small amount at a time tossing with a fork until all flour is moistened and pastry cleans side of bowl. (If dough is still dry, add a small amount additional ice water. Flour differs in absorption.) Gather pastry into a ball. Divide in half. Shape each half into a flattened round on lightly floured pastry cloth. Roll pastry with floured stockinet-covered rolling pin. Fold rolled pastry in half. Place in pie pan and unfold and ease into corners of plate pressing firmly against bottom and side. Makes 1 double crust pie or 2 single shells. Baked shells: prick bottom and sides with fork. Place another pan same size on crust. Bake at 475° about 8 minutes. Remove top pie pan and continue baking until light brown. Cool.

100% WHOLE WHEAT PIE CRUST

2½ c. stone ground whole wheat flour
½ tsp. salt
1 egg
2 tblsp. vinegar
⅔ c. shortening
4 tblsp. cold water
¾ tsp. baking powder

Sift dry ingredients, cut in shortening with pastry blender. Mix egg, vinegar, and water, then add to dry ingredients, mixing with fork. Roll out and bake filled with choice of fillings. This is a tender crust, but not as flaky as when white flour is used.

CAKES

AMAZIN' RAISIN CAKE

1¼ c. all-purpose flour
1½ c. stone ground whole wheat flour
2 c. sugar
1 c. mayonnaise
⅓ c. milk
2 eggs
3 tsp. baking soda
1½ tsp. cinnamon
½ tsp. nutmeg
½ tsp. salt
¼ tsp. cloves
3 c. peeled, chopped apples
1 c. raisins
½ c. chopped walnuts
Whipped cream or frosting

In large bowl with mixer at low speed beat together flour, sugar, mayonnaise, milk, eggs, baking soda, cinnamon, nutmeg, salt and cloves. Beat 2 minutes (batter will be thick), then stir in apples, raisins and nuts. Spoon batter into 9 x 13-inch pan. Bake in 350° oven 1 hour and 10 minutes, or until done. Frost with whipped cream or frosting of choice.

APPLE HARVEST CAKE

1¼ c. all-purpose flour
1 c. stone ground whole wheat flour
1 c. sugar
¾ c. firmly packed brown sugar
1 tblsp. cinnamon
2 tsp. baking powder
1 tsp. salt
½ tsp. soda
¾ c. cooking oil
1 tsp. vanilla
3 eggs
2 c. apples, peeled, finely choppped
¾ to 1 c. chopped nuts
½ c. powdered sugar
¼ tsp. vanilla
2 to 3 tsp. milk

Preheat oven to 325°. Generously grease and flour 12-cup fluted tube pan. Lightly spoon flour in measuring cup; level off. In large mixer bowl, blend sugars, cinnamon, baking powder, salt, soda, oil, vanilla and eggs until moistened; beat 3 minutes at medium speed. By hand, stir in apples and nuts. Pour into prepared 9 x 13-inch pan. Bake 50 to 65 minutes until toothpick inserted in center comes out clean. Cool upright in pan 15 minutes. Turn onto serving plate. Cool. In small bowl, blend glaze ingredients (powdered sugar, ¼ tsp. vanilla, milk) until smooth; spoon over cake.

APPLESAUCE RAISIN CAKE

1¼ c. all-purpose flour
1½ c. stone ground whole wheat flour
1 tsp. baking soda
1 tsp. ground cinnamon
½ c. butter OR margarine
1 beaten egg
1 c. molasses
1 8½-oz. can applesauce (1 cup)
½ c. raisins
¾ c. sifted powdered sugar
4 tsp. lemon juice

In a medium bowl combine flours, soda, cinnamon, and ½ tsp. salt. Cut in butter or margarine to resemble coarse crumbs. In another small bowl stir together egg, molasses, and applesauce; blend into flour mixture just till moistened. Stir in raisins. Turn into a greased and floured 9 x 9 x 2-inch baking pan. Bake at 350° for 40 to 45 minutes or till cake tests done. Cool in pan 15 minutes. Combine powdered sugar and lemon juice; spread over warm cake. Serve warm or cool.

CARROT CAKE

⅔ c. vegetable oil
1 c. sugar
2 c. stone ground whole wheat flour
1¼ tsp. baking soda
1¼ tsp. baking powder
1 tsp. salt
1 tsp. cinnamon
½ tsp. nutmeg
1 c. plain yogurt OR sour milk OR buttermilk
2½ c. grated carrots
½ c. chopped walnuts

In large bowl mix all ingredients in order given. Pour into greased pan (9 x 9-inch size). Bake at 350° for 40-50 minutes or until toothpick inserted in center comes out clean
Icing: 1 3-oz. pkg. cream cheese, 1 c. confectioner's sugar, ¾ tsp. vanilla. Cream together well. Add a little milk as needed to make a spreadable consistency. Makes nine 3 x 3-inch servings.

CHOCOLATE SHEET CAKE

2 c. stone ground whole wheat flour
1 c. sugar
¾ c. honey
4 tblsp. cocoa
½ c. vegetable oil
½ c. butter
1 c. water
2 eggs, beaten
1 c. buttermilk
1 tsp. soda
1 tsp. vanilla

Icing:
⅓ c. milk
¼ c. cocoa
½ c. butter
2 - 3 c. confectioner's sugar
1 c. nuts

Heat oven to 350°. Mix flour, sugar and honey. Bring cocoa, oil, butter and water to a boil. Pour over flour mixture and mix well. Add the remaining ingredients and pour into a greased and floured 10 x 15 x 1-inch jelly roll pan. Bake 25-30 minutes. Icing: boil milk, cocoa and butter. Remove from heat and add sugar and nuts. Frost cake while still warm.

CHOCOLATE ZUCCHINI CAKE

1¾ c. sugar
2 c. finely ground zucchini
½ c. soft oleo
½ c. oil
2 eggs
1 tsp. vanilla
½ tsp. salt
½ c. sour milk
1 c. all-purpose flour
1¼ c. stone ground whole wheat flour
4 tblsp. cocoa
⅛ tsp. soda
½ tsp. baking powder
½ tsp. cinnamon
½ c. chocolate chips
½ c. nuts

Cream oil, sugar and egg. Add zucchini, vanilla and milk. Mix in dry ingredients and fold in chocolate chips and nuts. Bake 45 minutes at 325°.

CHOCOLATE ZUCCHINI CAKE

1¼ c. stone ground whole wheat flour
1 c. all-purpose flour
½ c. cocoa
2½ tsp. baking powder
1½ tsp soda
1 tsp. salt
1 tsp. cinnamon
¾ c. margarine
2 c. sugar
3 eggs
2 tsp. vanilla
2 tsp. grated orange peel
2 c. chopped zucchini
½ c. milk
1 c. chopped nuts

Combine flour, cocoa, baking powder, soda, salt and cinnamon. Set aside. Beat together the margarine and sugar. Add eggs, one at a time; beat well. Stir in vanilla, orange peel and zucchini. Alternately, stir in dry ingredients and milk, including nuts in last addition. Bake at 350° in 10-inch tube or bundt pan for 1 hour or until done, or in three 8-inch layer pans. Layers take about 30 minutes.

COCOA-APPLE CAKE

1 c. butter OR margarine, softened
2 c. sugar
3 eggs
1¼ c. all-purpose flour
1¼ c. stone ground whole wheat flour
2 tblsp. cocoa
1 tsp. baking soda
1 tsp. ground cinnamon
1 tsp. ground allspice
½ c. water
2 c. finely chopped apples
1 c. finely chopped pecans
½ c. semisweet chocolate morsels
1 tsp. vanilla extract
Powdered sugar

Cream butter; gradually add sugar, beating well at medium speed of an electric mixer. Add eggs, one at a time, beating after each addition. Combine flour, cocoa, soda, cinnamon, and allspice; mix well. Add to creamed mixture alternately with water, beginning and ending with flour mixture. Mix just until blended after each addition. Stir in remaining ingredients except powdered sugar, mixing just until blended. Pour batter into a greased and floured 10-inch bundt pan. Bake at 325° for 60 to 65 minutes or until a wood pick inserted in center comes out clean. Cool in pan 10 to 15 minutes; remove from pan, and cool completely on a wire rack. Sprinkle cake with powdered sugar.

DUTCH APPLE CAKE

¾ c. stone ground whole wheat flour
½ c. all-purpose flour
2¼ tsp. baking powder
¼ tsp. salt
1 tblsp. sugar
4 tblsp. shortening
1 egg
½ c. milk
Cooking apples, peeled, cored and cut in eighths

Stir flours with baking powder, salt and sugar. Cut shortening in very fine. Beat egg and combine with milk; gradually add to flour mixture. Spread dough over bottom of well greased baking dish. Arrange apples over dough in overlapping rows, pointed edge down. Sprinkle with 2 tblsp. sugar and 1 tsp. cinnamon. Dot with butter. Bake at 350° until apples are tender, about 30 minutes. Serve hot with cream or a lemon sauce.

FRESH APPLE WALNUT CAKE

1 c. butter OR margarine
2 c. sugar
3 eggs
1¼ c. all-purpose flour
1½ c. stone ground whole wheat flour
3 tblsp. cornstarch
1½ tsp. baking soda
½ tsp. salt
1 tsp. cinnamon
¼ tsp. mace
3 c. chopped apples
2 c. chopped walnuts

Cream butter and sugar until fluffy. Add eggs, 1 at a time, beating well after each addition. Mix and sift flour, cornstarch, baking soda, salt, cinnamon and mace; add gradually. Stir in vanilla, apples and walnuts. Batter will be stiff. Spoon into greased and floured 10-inch tube pan. Bake at 325° for 1½ hours. Let cool in pan 10 minutes. Remove from rack.

GERMAN APPLE CAKE

2 c. sugar
1 c. all-purpose flour
1 c. stone ground whole wheat flour
2 tsp. cinnamon
1 tsp. baking soda
½ tsp. salt
4 c. pared, sliced apples
½ c. chopped nuts
2 eggs
1 c. salad oil
1 tsp. vanilla
Coffee glaze, optional

Preheat oven to 350°. Grease and flour 9 x 13-inch pan. Stir together sugar, flour, cinnamon, soda and salt. Stir in apples and nuts. Beat eggs with oil and vanilla; pour into dry ingredients and stir; do not use mixer. Pour into prepared pan and bake in preheated oven 45 to 60 minutes, or until cake is browned and tests done. (Baking time will vary depending on juiciness of apples used.) Makes 9 to 12 servings. Coffee Glaze: In large glass measuring pitcher, combine about 1 tblsp. instant coffee and ⅛ to ¼ cup water. Heat to boiling in microwave; add about 1 tblsp. butter or margarine. Stir confectioner's sugar into hot liquid to thin-glaze consistency; stir in a few drops of vanilla. Drizzle and spread over cooled cake.

GOLDEN WHOLE WHEAT ANGEL FOOD CAKE

8 egg yolks
1 c. cold water
2 c. sugar
2 c. stone ground whole wheat flour
½ c. cornstarch
½ tsp. salt
1½ tsp. vanilla
8 egg whites
1 tsp. cream of tartar

Beat egg yolks until light colored. Add cold water and beat for 2 minutes. Add sugar and blend. Combine flour, cornstarch and salt and sift 3 or 4 times. Add to egg yolk mixture. Beat 3 or 4 minutes. Add vanilla. Beat egg whites with cream of tarter until very stiff. Fold into egg yolk mixture. Mix evenly. Bake in angel food cake pan 1¼ hours at 325°. Invert pan on pop bottle to cool (Fill cake pan ¾ full. Bake rest in loaf pan.) Optional: 1 tsp. **each** cinnamon, allspice, cloves. Mix with dry ingredients.

HONEY APPLESAUCE CAKE

½ c. shortening
1 c. honey
2 eggs
2¼ c. stone ground whole wheat flour
⅓ c. nonfat dry milk powder
1 tsp. baking soda
1 tsp. baking powder
½ tsp. salt
½ tsp. ground cinnamon
¼ tsp. ground cloves
1 c. applesauce
1 c. raisins
1 c. chopped walnuts
1 c. powdered sugar
¼ tsp. vanilla
1-2 tblsp. milk

In a large mixer bowl beat shortening with honey on high speed of electric mixer till light and fluffy. Beat in eggs, one at a time, beating well after each addition. In another bowl stir together flour, nonfat dry milk powder, soda, baking powder, salt, cinnamon, and cloves; add to creamed mixture alternately with applesauce, beating well after each addition. Fold in raisins and nuts. Pour batter into a greased and floured 10-inch tube or 12-cup fluted tube pan. Bake in a 325° oven for 35 to 45 minutes or till cake tests done. Immediately turn out onto wire rack; cool. In a small bowl stir together powdered sugar, vanilla, and enough milk to make a glaze consistency. Drizzle over cooled cake.

HONEY OATMEAL CAKE

1 c. quick-cooking oats
1 stick butter OR margarine
1¼ c. boiling water
1½ c. honey
1 tsp. vanilla
2 eggs, beaten
1¾ c. stone ground whole wheat flour
1 tsp. baking powder
¾ tsp. salt
1 tsp. cinnamon
¼ tsp. nutmeg

Put oats, butter and boiling water in large mixing bowl, and let stand for 20 minutes. Add honey, vanilla and eggs. Sift together all dry ingredients and then add to first mixture. Pour into greased and floured 9 x 13 x 2-inch cake pan. Bake 30 to 40 minutes in a 350° oven. Excellent plain or frosted with coconut-brown sugar frosting. Serves 10 to 12.

HONEY WHOLE-WHEAT ANGEL FOOD CAKE

6 eggs, separated
1 tsp. cream of tartar
1 c. honey
1¾ c. stone ground whole wheat flour
⅓ c. orange juice
½ tsp. cloves OR pumpkin pie spice
1 tsp. vanilla
Honey Raspberry Sauce:
2 c. raspberries
½ c. water
½ c. honey
2 tblsp. cornstarch

Cake: Beat egg whites and cream of tarter until very stiff peaks form; set aside. In large mixer bowl beat honey smooth; add egg yolks one at a time, beating after each until smooth. Blend in flour, orange juice, spice and vanilla. Fold in egg whites until blended. Pour into ungreased 10-inch tube pan. Bake at 325° for 50 to 60 minutes or until done. Cool on rack upside down. Sauce: Combine ingredients in saucepan. Cook stirring until mixture boils and thickens; cool. Remove cake from pan to cake plate. Pour sauce over top of cake, allowing it to drizzle down sides of cake. Top or serve cake with whipped cream.

OATMEAL CAKE

1 c. oatmeal
1½ c. boiling water
1½ c. stone ground whole wheat flour
½ tsp. salt
2 tsp. baking powder
1 tsp. cinnamon
¾ tsp. nutmeg
½ c. margarine OR butter
3 eggs
1⅓ c. packed brown sugar
1 tsp. vanilla

Pour boiling water over oatmeal and set aside. Sift dry ingredients together. Cream butter with sugar and vanilla and blend in eggs one at a time. Stir in dry ingredients alternately with oat mixture and beat well. Pour into greased, floured 9 x 13-inch pan and bake 35 to 40 minutes at 350°. One cup raisins may be added for variety. Frost with favorite frosting.

PEACH-COCONUT CAKE

2 eggs, beaten
½ c. vegetable oil
½ pint plain yogurt
1 tsp. vanilla
1 c. stone ground whole wheat flour
1 c. all-purpose flour
½ c. sugar
1 tsp. baking soda
1 c. shredded coconut
2 fresh peaches, sliced (1½ cups)
1 tblsp. cinnamon
Apricot jam, sieved (optional)

Beat eggs with oil, yogurt and vanilla until smooth. Combine flours, sugar, soda and coconut. Mix into egg mixture to form batter. Turn about one third of batter into greased 6-cup ring mold. Mix peaches with cinnamon and place half of the mixture on top of batter. Layer with another third of batter, then remaining peaches. Top with remaining batter. Bake in 350° oven 40 to 50 minutes or until pick inserted in center comes out dry. (Fruit layers should remain moist.) Let cool 10 minutes in pan, then invert and cool on rack. Refrigerate cake, if any remains after the first day, to prevent browning of the fruit. Serve with glaze of sieved apricot jam for dessert or, if preferred, omit glaze and serve for breakfast or brunch. Makes 6 to 8 servings.

PINEAPPLE CAKE

2 eggs, well beaten
1 c. sugar
1 tsp. vanilla
1 c. crushed pineapple (drained)
1 c. chopped nuts
¾ c. stone ground whole wheat flour
1 tsp. baking powder
¼ tsp. salt

Combine and mix eggs, sugar and vanilla. Add pineapple and nuts. Fold in flour, baking powder and salt. Bake 30 to 35 minutes at 325° in 9 x 9-inch greased and floured cake pan. Cool. Top with topping poured over each slice of cake as you serve it. Add whipped cream and a cherry for a final touch.
Topping: Blend and boil for 3 minutes, ¼ c. butter, 1 tblsp. flour, 1 c. brown sugar, packed, ¼ c. pineapple juice, ¼ c. water. Then take off heat and add, stirring vigorously, 1 well beaten egg. Add 1 tsp. salt.

PUMPKIN WALNUT CAKE

1¼ c. all-purpose flour
1½ c. stone ground whole wheat flour
2 c. (1 lb. can) pumpkin
2 c. sugar
1½ c. vegetable oil
1 c. coarsely chopped walnuts
4 large eggs
3½ tsp. cinnamon
2 tsp. baking powder
2 tsp. baking soda
1 tsp. salt
Powdered sugar, optional

Beat eggs well. Gradually add sugar, beating until thick. Beating constantly, pour in oil. With mixer at low speed, blend in sifted dry ingredients, alternately with pumpkin; beginning and ending with dry ingredients. Beat until smooth after each addition. Stir in walnuts. Bake in ungreased angel food pan in 350° oven for approximately 1 hour and 10 minutes or until a toothpick comes out clean. Cool completely. Remove from pan and sprinkle with powdered sugar or top with whipped cream. This cake freezes very well.

SPICEY WHOLE-WHEAT CAKE

1 c. stone ground whole wheat flour
½ c. all-purpose flour
1 c. sugar
1 tsp. baking soda
1 tsp. cinnamon
½ tsp. each salt and nutmeg
¼ tsp. cloves
½ c. whole-bran cereal
1 c. cold, strong coffee
¼ c. oil
1 tblsp. vinegar
1 tsp. vanilla

Mix well flours, sugar, baking soda, spices and salt; set aside. In pan, mix well cereal and coffee; let stand 2 minutes or until most of liquid is absorbed. Stir in oil, vinegar and vanilla, then stir in flour mixture until smooth. Bake for 40 minutes in 350° oven in an 8 x 8 x 2-inch baking pan, or until pick inserted in center comes out clean. Cool completely in pan on rack.

SUNKEN APPLE CAKE

1 c. unsalted butter
1 c. sugar
4 egg yolks
1 tsp. vanilla
Grated peel of ½ lemon
Pinch of salt
1 c. all-purpose flour
1 c. stone ground whole wheat flour
½ c. sifted cornstarch
3 tsp. baking powder
4 egg whites, beaten

Topping:
1½ lb. apples, peeled, cored and sliced
1 tblsp. sugar
2 tsp. lemon juice
1 c. finely ground filberts or almonds
3 tblsp. sugar
¼ c. raisins, optional

Preheat oven to 375°. Butter and flour a 10-inch springform pan. Cream butter, sugar and egg yolks. Add vanilla, lemon peel, and salt. Stir in flour, cornstarch and baking powder. Fold in beaten egg whites. Pour into prepared pan.
Topping: Sprinkle apple slices with sugar and lemon juice; stir. Arrange apple slices so that they overlap each on the next, forming circles on top of the cake. Press the slices down slighty, inserting raisins between them. Sprinkle combined nuts and sugar evenly over all. Place in lower middle of oven and bake 50 minutes or until done. Cool 10 minutes in the pan before turning onto cake rack to cool.

WHOLE WHEAT CARROT CAKE

1 c. cooking oil
1 c. sugar
1 c. packed brown sugar
1 tsp. vanilla
4 eggs
2 c. stone ground whole wheat flour
⅓ c. nonfat dry milk powder
1 tsp. baking soda
1 tsp. baking powder
2 tsp. ground cinnamon
3 c. finely shredded carrots
1 c. chopped walnuts
Powdered sugar

In a large mixer bowl blend oil and sugars on low speed of electric mixer until mixed. Add vanilla; beat in eggs, one at a time, beating well after each addition. In another bowl stir together flour, milk powder, baking soda, baking powder, 1 tsp. salt and cinnamon. Add to egg mixture till well blended. By hand, stir in carrots and walnuts. Pour batter into a greased and floured 10-inch tube pan or 12-cup fluted tube pan. Bake in 350° oven for 50 to 60 minutes or till cake tests done. Cool in pan; invert onto serving plate. Sprinkle sifted powdered sugar on top.

WHOLE WHEAT DATE CAKE

2 c. chopped dates
1 c. water
1 c. butter OR margarine
1½ c. all-purpose flour
1½ c. stone ground whole wheat flour
2 tblsp. baking powder
1 tsp. salt
1 tsp. cinnamon
½ tsp. nutmeg
1 c. brown sugar, packed
4 eggs, beaten
1 tblsp. vanilla

Combine dates, water and butter and cook over low heat until caramel-colored and thickened. Refrigerate to cool. Sift together dry ingredients and add to date mixture along with eggs and vanilla. Stir until well mixed and pour into a 10-inch tube pan (not bundt). Bake at 350° for 50 to 55 minutes. Cool for 10 minutes and remove from pan. Cool completely on a rack. Serve with whipped topping mixed with applesauce.

WHOLE WHEAT ANGEL FOOD CAKE

¾ c. stone ground whole wheat flour
¼ c. cornstarch
1½ c. sugar
12 large egg whites
½ tsp. salt
½ tsp. cream of tarter
1 tsp. vanilla or almond extract

Heat oven to 375°. Combine flour, cornstarch, and ¾ c. sugar in a small bowl. Stir until thoroughly mixed. Set aside. Separate egg whites being sure there is no yolk mixed with the whites. Add salt and cream of tarter. Whip until whites will stand in peaks. Gradually add remaining ¾ c. sugar and flavoring. Sprinkle ⅓ of the flour mixture over the beaten whites; fold in carefully. Repeat twice with the remaining flour mixture, folding the last one-third only until it is thoroughly mixed. Pour into angel food cake pan; bake 40 minutes. Invert pan to cool.

Knowledge may be impersonal.
Wisdom is not.

WHOLE-WHEAT ANGEL FOOD CAKE

¾ c. stone ground whole wheat flour
1½ c. sugar
1⅔ c. (about 13) egg whites
1½ tsp. cream of tarter
1½ tsp. orange extract
½ tsp. salt
Orange Chiffon Cream Filling (recipe follows)
¾ c. whipping cream
2 tblsp. confectioner's sugar
2 navel oranges, peeled and sectioned

Sift flour and ¾ c. sugar 4 times onto a large piece of waxed paper. In a 5-qt. bowl beat egg whites with electric mxer on medium speed about 1 minute until foamy. Add cream of tarter, orange extract and salt. Gradually beat in the remaining ¾ c. sugar until egg whites are moist but stiff and adhere to side of bowl when bowl is tipped. Sift about one-third the flour mixture at a time over egg whites. With mixer on lowest speed, fold in each addition until blended, scraping down sides of bowl with rubber spatula if necessary. Pour into an ungreased 10-inch tube pan. Cut through batter with a knife to burst any air pockets; smooth top. Bake on middle rack of preheated 350° oven 40 to 50 minutes until pick inserted near center of cake comes out clean. Invert pan on supports around rim or over the neck of a bottle. Cool at least 2 hours or overnight. Turn cake right side up. Run long, thin-bladed knife or spatula between cake and edges of pan to loosen. Remove from pan; turn right side up. Slice cake horizontally into 4 layers. To assemble, place 1 layer on serving plate, spread with one-third the Orange Chiffon Cream Filling. Repeat with remaining layers and filling topping with last layer. Whip cream and confectioner's sugar until it holds its shape and is of spreading consistency. Spread on top layer; chill. Garnish with orange sections just before serving. Makes 12 servings.

ORANGE CHIFFON CREAM FILLING

1 envelope unflavored gelatin
3 tblsp. cold water
1 c. milk
Yolks from 4 large eggs
⅓ c. sugar
¼ tsp. salt
1 tblsp. grated orange peel
1 tsp. vanilla
2 tblsp. orange juice concentrate
1 c. whipping cream, whipped

Sprinkle gelatin over cold water in a small bowl; stir; let stand about 5 mintues to soften. Meanwhile in a heavy saucepan, heat the milk over medium heat until small bubbles appear around edges. In a medium bowl beat egg yolks with electric mixer until pale yellow. Gradually beat in sugar, then the salt. Add the hot milk stirring constantly. Stir in gelatin and orange peel. Return to saucepan and cook over medium heat until slightly thickened, stirring constantly. Remove from heat; pour into bowl; let stand about 2 minutes to cool slightly. Chill, stirring occassionally. Stir in vanilla and orange juice. Gently fold in whipped cream until blended. Refrigerate about 1 hour until the mixture thickens and sets, stirring occassionally. Makes about 3 cups.

ZUCCHINI CAKE

1½ c. oil
4 eggs
2 c. sugar
2 c. stone ground whole wheat flour
1 c. all-purpose flour
2 tsp. baking powder
1 tsp. soda
1 tsp. cinnamon
1 tsp. salt
3 c. grated zucchini
1 c. nuts (optional)

Beat eggs, add sugar and oil. Mix well. Add dry ingredients and mix well. Stir in greated zucchini and nuts. Pour into a greased 9 x 13-inch pan and bake at 300° for one hour or until done.

ZUCCHINI TEACAKE

1 c. all-purpose flour
2 c. stone ground whole wheat flour
½ c. cornstarch
4 tsp. baking powder
1½ tsp. salt
1 tsp. cinnamon
1½ tsp. baking soda
2 c. sugar
1 c. vegetable oil
1 tsp. vanilla
4 large eggs
2 c. grated zucchini
1 c. chopped walnuts
1 c. dried currants

Grease six mini loaf pans or two regular loaf pans. On waxed paper stir together the flour, baking powder, baking soda, salt and cinnamon. In a large mixing bowl beat together the sugar, oil, and vanilla until combined; thoroughly beat in the eggs one at a time. Stir in the flour mixture in several additions alternately with the zucchini. Stir in the walnuts and currants. Turn into the prepared pans—they will be about ½ full. Bake in a preheated 350° oven until a cake tester inserted in center comes out clean—45 minutes for mini loaves, 55 to 60 minutes for large. Loosen edges and turn out on wire racks; turn right side up; cool completely.

COOKIES & BARS

APPLESAUCE JUMBLES

1½ c. stone ground whole wheat flour
1¼ c. all-purpose flour
1¼ c. brown sugar (packed)
1 tsp. salt
½ tsp. baking soda
1 tsp. cinnamon
¼ tsp. cloves
¾ c. applesauce
¼ c. shortening
2 eggs
1 tsp. vanilla
1 c. raisins
1 c. chopped nuts, if desired
Browned butter glaze (below)

Mix thoroughly all ingredients except glaze. If dough is soft, cover and chill. Heat oven to 375°. Drop dough by level tblsp. 2 inches apart onto ungreased baking sheet. Bake 10 minutes or until almost no imprint remains when touched with finger. Immediately remove from baking sheet; cool. Spread with browned butter glaze. Makes 4½ to 5 dozen.

Browned Butter Glaze: Heat ⅓ c. butter or margarine over low heat until golden brown. Remove from heat, blend in 2 c. confectioner's sugar and 1½ tsp. vanilla. Stir in 2 to 4 tblsp. hot water until of spreading consistency.

BANANA-OATMEAL COOKIES

1 c. sugar
¾ c. shortening
1 egg, beaten
1 c. mashed banana
¼ tsp. nutmeg
¾ tsp. cinnamon
¾ c. stone ground whole wheat flour
¾ c. flour
½ tsp. soda
¼ tsp. salt
1¾ c. uncooked, quick oats
1 c. chopped nuts
1 pkg. (6-oz.) chocolate chips (optional)

Preheat oven to 350°. Cream sugar and shortening; add egg and mashed banana. Sift together nutmeg, cinnamon, flour, soda and salt. Stir into creamed mixture. Mix together nuts and oatmeal and stir in. Drop by teaspoonfuls onto greased cookie sheets and bake 15 minutes. Makes about 6 dozen cookes.

CAROB COOKIES

½ c. margarine OR butter
¾ c. firmly packed brown sugar
1 egg
1 c. plus 2 tblsp. stone ground whole wheat flour
3 tblsp. carob powder
½ tsp. baking powder
¼ tsp. soda
¼ tsp. salt
1½ tsp. vanilla
½ c. hulled sunflower seeds
½ c. raisins

Cream the margarine and sugar; add the egg. Sift dry ingredients together and add to the creamed mixture, mixing thoroughly. Stir in the vanilla, sunflower seeds and raisins. Drop by teaspoonful on greased cookie sheet. Bake 10 minutes at 375°. Makes 3½ dozen cookies. A nutritious cookie, chocolate flavored for those allergic to chocolate; with the taste of the finest chocolate to satisfy the chocolate lover.

CARROT COOKIES

½ c. butter OR margarine, at room temperature
½ c. sugar
1 egg
¼ c. orange marmalade
1½ c. grated carrots (about 2 medium carrots)
⅔ c. stone ground whole wheat flour
⅔ c. all-purpose flour
½ tsp. salt
1 tsp. baking powder
½ c. chopped raisins or currants

Preheat oven to 350°. Lightly grease 2 baking sheets. In medium bowl, beat butter or margarine and sugar until light and fluffy. Beat in egg until blended; stir in marmalade and carrots. Sift flour, salt and baking powder over butter mixture. Fold in dry ingredients and raisins or currants. Drop rounded teaspoons of mixture, about 1½ inches apart, on greased baking sheets. Bake in preheated oven 12 to 15 minutes or until golden around edges. Remove from baking sheets; cool on wire racks. Makes about 48 cookies.

CARROT-OATMEAL COOKIES

⅞ c. stone ground whole wheat flour
¼ c. nonfat dry milk powder
1½ tsp. baking powder
½ tsp. baking soda
½ tsp. salt
¼ tsp. ground nutmeg
¼ tsp. ground cinnamon
⅓ c. shortening
⅓ c. brown sugar
½ c. molasses
1 egg
1 c. shredded carrots
1 tsp. vanilla
½ c. raisins (optional)
1¾ c. quick cooking rolled oats

Combine flour, dry milk, baking powder, baking soda, salt, nutmeg and cinnamon. Cream together shortening, sugar and molasses; add egg, then dry ingredients. Stir until well blended. Add carrots, vanilla, raisins and oats; mix well. Drop by teaspoonfuls onto an ungreased cookie sheet. Bake in a preheated 375° oven for 10-12 minutes, until lightly brown. Remove cookies and cool on wire racks. Makes 4 dozen.

CHEESECAKE COOKIES

⅓ c. butter OR margarine
⅓ c. brown sugar
1 c. stone ground whole wheat flour
½ c. chopped walnuts, OR toasted sesame seeds, OR roasted sunflower seeds
¼ c. honey
8 oz. cream cheese
1 egg
2 tblsp. milk
1 tblsp. lemon juice
Grated peel of 1 lemon
½ tsp. vanilla extract
½ tsp. nutmeg (optional)

Garnish (optional):
Fruit slices: orange, apple, banana, strawberry; chopped nut meats: almonds, walnuts, brazil nuts

Blend together with pastry cutter to make a crumbly texture: whole wheat flour, brown sugar, and butter or margarine. Mix in chopped nuts or seeds. Reserve ½ c. for topping. Press remainder into oiled 8-inch square pan and bake at 350° for 12 - 15 minutes. Soften cream cheese with mixing spoon. Blend in honey. Blend in remaining ingredients and beat well. Spread over baked crust. Sprinkle reserved crumbs on top. Garnish with fruit slices and nut meats. Bake at 350° for 25 minutes. Cool and cut in 2-inch squares. Note: if using strawberries for garnish, place on cheesecake after baking.

COCOA FROSTEDS

1½ c. sugar
½ c. cocoa
¾ c. butter OR margarine, melted
1 egg
2 tsp. vanilla
¼ c. buttermilk
½ tsp. baking soda
¼ tsp. cream of tartar
1 c. all-purpose flour
1 c. stone ground whole wheat flour
1 tsp. cinnamon
1 tsp. cloves
1 tsp. nutmeg
Pecan halves

Cream together sugar, cocoa and melted shortening. Add egg and vanilla and beat well. Dissolve soda and cream of tartar in buttermilk (will foam up in cup). Stir into creamed mixture. Mix or sift dry ingredients together and stir into mixture. Chill dough. Roll into balls the size of large walnuts and place 2 inches apart on ungreased cookie sheet. Bake in preheated 350° oven for 8 minutes. Immediately remove from baking sheet to cooling rack. Frost when cool and put a pecan half on each cookie. Yield - 4 dozen large cookies.

COCOA SPICE FROSTING

1/3 c. soft butter OR margarine
1/3 c. cocoa
2 c. powdered sugar
1/2 tsp. cinnamon
1/4 tsp. nutmeg
1 1/2 tsp. vanilla
About 2 tblsp. milk

Mix butter and cocoa thoroughly. Blend spices with sugar and add to butter mixture. Stir in vanilla and enough milk to make a smooth, spreadable frosting.

DATE BALLS

1 1/4 c. sugar
2 eggs
1/2 c. stone ground whole wheat flour
1 tsp. baking powder
1/4 tsp. salt
1 c. pitted dates, chopped
1 c. chopped pecans

Preheat oven to 350°. Grease an 8-inch square pan; set aside. In mixing bowl with electric mixer at medium speed, beat 1 cup sugar and eggs until fluffy. In another bowl combine flour, baking powder and salt. Add to the sugar mixture and blend. Fold in dates and pecans. Spoon batter into prepared pan. Bake for 30 minutes. Remove from oven and stir immediately with a wooden spoon; let cool completely. Roll into 1-inch balls, then in remaining 1/4 cup sugar. Makes 3 dozen.

DIABETIC COOKIE

1 pkg. dry yeast
1/2 c. warm water
1 c. vegetable oil
1 c. stone ground whole wheat flour
1 1/4 c. all-purpose flour
1 tsp. vanilla
Diabetic sugar

Preheat oven to 375°. Soften yeast in warm water. Set aside for 5 minutes. Beat at high speed, the oil, then add at low speed the flour with yeast mixture and vanilla. Shape into small balls and roll in diabetic sugar. Bake for 20 minutes on ungreased cookie sheet.

GOLDEN HONEY CARROT COOKIES

- 1/4 c. shortening
- 1/4 c. butter OR margarine
- 1/2 c. sugar
- 1/4 c. honey
- 1 egg
- 1 tsp. vanilla extract
- 1/2 tsp. lemon extract
- 1 1/4 c. grated raw carrots
- 1 c. all-purpose flour
- 1 c. stone ground whole wheat flour
- 2 1/2 tsp. baking powder
- 1/2 tsp. salt

Beat together shortening, butter, and sugar until light and fluffy. Add honey, egg and extracts; beat well. Stir in carrots. Thoroughly mix together flours, baking powder and salt; stir into shortening mixture. Drop by rounded teaspoons onto greased baking sheets. Bake at 350° for 10 minutes or until golden brown. Makes 3 1/2 dozen cookies.

GOOD-FOR-YOU COOKIES

- 1 c. raisins
- 1 c. dried apricots
- 1/2 c. non-fat dry milk
- 1/2 tsp. soda
- 3/4 tsp. salt
- 1/4 tsp. baking powder
- 3/4 c. stone ground whole wheat flour
- 1/3 c. wheat germ
- 1/2 c. butter
- 1 c. brown sugar
- 1 egg
- 1 tsp. vanilla
- 3 tblsp. milk
- 1/3 c. raw sunflower seeds
- 1 c. oatmeal

Coarsely chop apricots and raisins. Mix dry milk, baking powder, soda and salt; stir in wheat germ and flour. Cream butter; add sugar and beat until fluffy. Add egg; beat well. Add vanilla; beat well. Add flour mixture and milk alternately. Add seeds and oats, work in fruit. Drop by teaspoonfuls onto greased cookie sheets. Bake in 375° oven 7 minutes. Cool 5 minutes on cookie sheet before removing to cooling rack or waxed paper. Makes about 3 dozen cookies.

GRANDMA'S ZUCCHINI COOKIES

- 3/4 c. butter OR margarine
- 1 1/2 c. sugar
- 1 egg
- 1 tsp. vanilla
- 1 1/2 c. grated zucchini
- 1 1/4 c. all-purpose flour
- 1 1/4 c. stone ground whole wheat flour
- 2 tsp. baking powder
- 1 tsp. cinnamon
- 1/2 tsp. salt
- 1 c. coarsely chopped whole natural almonds
- 1 c. (6-oz.) semi-sweet chocolate pieces

Cream butter with sugar. Beat in egg and vanilla; mix in zucchini. Stir in flour, baking powder, cinnamon and salt. Stir in almonds and chocolate pieces. Drop by heaping teaspoonfuls onto greased cookie sheet. Bake at 350° for 15 minutes or until lightly browned. Cool on wire rack. Sieve powdered sugar over cookies, if desired. Makes about 4 dozen.

GRANDMOTHER'S OATMEAL-PRUNE COOKIES

1¼ c. quick-cooking OR old-fashioned oats, uncooked
¾ c. stone ground whole wheat flour
¾ c. packed brown sugar
⅓ c. shortening
1 tsp. vanilla extract
¾ tsp. baking soda
¼ tsp. salt
1 egg
¾ c. pitted prunes, chopped

Into large bowl, measure all ingredients except prunes. With mixer at low speed, beat ingredients until well blended, occasionally scraping bowl with rubber spatula. Stir in prunes. Preheat oven to 350°. Drop mixture by heaping tablespoonfuls, about 2 inches apart, onto ungreased large cookie sheets. Bake cookies 15 minutes or until golden. Carefully remove cookies to wire racks to cool. Store cookies in tightly covered container. Makes about 2 dozen cookies.

HONEY / WHEAT NUGGETS

⅓ c. peanut butter
⅓ c. margarine
½ c. honey
1 egg
1 tsp. vanilla
⅔ c. stone ground whole wheat flour
¼ tsp. salt
½ tsp. soda
1 tsp. baking powder
½ c. non-fat dry milk
½ c. wheat germ
1 c. quick cooking oats
1 c. raisins
½ c. chopped nuts
½ c. coconut
½ c. tiny chocolate chips

Cream peanut butter, margarine, and honey. Add egg and vanilla, mix well. Add whole wheat flour, salt, soda and baking powder. Stir in powdered milk, wheat germ, and oats. Stir in remaining ingredients separately. Drop balls, 1 inch in diameter, onto greased cookie sheet. Bake 8-10 minutes at 350°.

A good example is twice as valuable as good advice.

MAPLE SCOTCH COOKIES

1 c. stone ground whole wheat flour
1 c. all-purpose flour
2 c. brown sugar
½ c. butter OR margarine
1½ tsp. soda
½ tsp. salt
1 egg, slightly beaten
2 tblsp. milk
½ tsp. maple extract
½ c. pecans, chopped
2 tblsp. instant chocolate mix

Preheat oven to 375°. Combine flour and brown sugar. Cut in shortening. Remove ¼ cup and add the chocolate mix for topping. Set aside. To the remainder add the soda and salt. Mix well. Blend in the egg, milk and maple extract. Mix at low speed to form soft dough. Stir in pecans. Shape into 1-inch balls and roll in the topping mixture. Place on greased cookie sheet and bake 12 to 14 minutes. Makes 4 dozen cookies.

NUT AND BUTTER COOKIES

½ c. butter
2 tblsp. honey
1 tsp. vanilla
¼ tsp. salt
1½ c. stone ground whole wheat flour
⅓ c. ground OR finely chopped walnuts
1 tsp. baking powder
½ c. strawberry preserves

Cream together butter and honey, then add vanilla, salt and remaining ingredients except the preserves. Mix well and form small walnut-sized balls of the dough. Make an indention in the center of each to fill with preserves after baking. Bake 12-15 minutes at 400° on an ungreased cookie sheet. Place on a wire rack to cool and fill with preserves while still warm.

OATMEAL COOKIES

1 c. shortening, soft
1 c. granulated sugar
1 c. firmly-packed brown sugar
2 eggs
1 tsp. vanilla
½ c. sifted all-purpose flour
½ c. stone ground whole wheat flour
1½ tsp. baking powder
½ tsp. salt
1 tsp. cinnamon
¼ tsp. nutmeg
3 c. quick cooking rolled oats, uncooked

Heat oven to moderate (350°). Place shortening, sugars, eggs and vanilla in mixing bowl; beat thoroughly. Sift together flour, baking powder, salt, cinnamon, and nutmeg; add to shortening mixture. Mix thoroughly. Stir in oats. Drop from teaspoon onto greased cookie sheets. Bake in preheated oven (350°) 12 to 15 minutes. For variety, add chopped nuts, raisins, chocolate chips, dates or coconut to the dough. Makes 6 dozen cookies.

OATMEAL CRISPS

1 c. shortening
1 c. brown sugar
1 c. granulated sugar
2 eggs, beaten
1 tsp. vanilla
¾ c. sifted all-purpose flour
¾ c. stone ground whole wheat flour
1 tsp. salt
1½ tsp. soda
3 c. quick oats, dry

Cream shortening and sugars. Add eggs and vanilla. Beat well. Add dry ingredients and mix well. Roll into walnut-size balls and place on ungreased cookie sheet. Pat each cookie once or twice with palm of hand to flattan slightly. Bake in 350° oven for 10 minutes. Makes 5 to 6 dozen cookies.

PEANUTTY OATMEAL TREATS

¾ c. stone ground whole wheat flour
1 c. all-purpose flour
1 tsp. baking soda
½ tsp. salt
½ c. butter OR margarine, softened
½ c. chunk-style peanut butter
1 c. granulated sugar
1 c. packed brown sugar
2 eggs
¼ c. milk
1 tsp. vanilla
2½ c. uncooked oats
½ c. each semisweet chocolate pieces and raisins

Stir together flour, soda and salt; set aside. In large bowl beat butter, peanut butter and sugars until creamy. Beat in eggs, milk and vanilla. Stir in flour mixture, then oats, chocolate pieces and raisins. Drop by rounded tablespoonfuls 3 inches apart on ungreased baking sheets. Bake in preheated 350° oven about 15 minutes or until lightly browned. Remove to rack to cool. Makes 42 cookies.

POWDERED ALMOND COOKIES

¾ c. margarine, softened
¼ c. powdered sugar
1 tblsp. cold water
1 tsp. almond extract
1 c. stone ground whole wheat flour
1 c. ground almonds
Dash of salt
1 c. powdered sugar
½ tsp. ground cinnamon

Cream margarine until light and fluffy. Stir in sugar, then water and almond extract; mix well. Add flour, almonds and salt. Shape into two long rolls, approximately 1½ inches in diameter. Wrap in foil or waxed paper and chill in refrigerator until hardened. Unwrap and slice into ¼-inch slices. Place on ungreased cookie sheet. Bake in 400° oven for 8 minutes or until lightly browned on top and bottom. While cookies are baking, stir cinnamon into sugar. Roll hot cookies in cinnamon/sugar mixture. Place on rack to cool. Makes 48 cookies.

RAISIN-OATMEAL COOKIES

⅔ c. butter OR margarine, softened
½ c. packed brown sugar
6 tblsp. granulated sugar
2 eggs
1 tsp. vanilla
1½ c. rolled oats
½ c. all-purpose flour
½ c. stone ground whole wheat flour
1 tsp. baking soda
½ tsp. salt
1 c. raisins

In large bowl, combine butter, eggs, sugars and vanilla. Beat with electric mixer about 2 minutes until mixture is fluffy. Combine oats, flour, baking soda and salt; add to beaten mixture. Beat until thoroughly blended. Stir in raisins. Drop by rounded teaspoonfuls 2 inches apart on greased baking sheets. Bake in 350° oven 10 to 12 minutes until lightly browned. Makes 3 dozen cookies.
Variations:
Add 1 cup of any combination of the following ingredients to the basic dough when blending in the raisins: grated carrots, shredded coconut, chopped nuts.
Substitute chocolate-covered raisins for plain raisins.
Substitute peanut butter for ½ of the butter in basic recipe.
For a bar cookie, add ½ cup each chopped nuts, coconut and mashed ripe banana to basic dough. Bake in greased 8-inch square pan 30 to 40 minutes, until golden. Cool; cut into squares.

SOFT OATMEAL COOKIES

1 c. sugar
1 c. shortening
2 c. cooked raisins
½ c. raisin water, cooled
1 tsp. vanilla
2 c. oatmeal
1 c. all-purpose flour
1 c. stone ground whole wheat flour
1 tsp. soda
1 tsp. salt

Combine sugar and shortening. Add raisin water, raisins and vanilla. Sift together flour, soda and salt; stir in oatmeal. Add flour-oatmeal mixture to creamed mixture. Drop by spoonfuls onto greased cookie sheet. Bake in 350° oven 10 minutes.

SPICY WHEAT COOKIES

½ c. plus 1 tblsp. unsalted sweet butter, softened
⅓ c. sugar
3 egg yolks
1 ⅓ c. all purpose flour
⅓ c. whole wheat flour
3 egg whites, at room temp.
1¼ c. sugar
1 c. finely ground almonds
1 tsp. ground cinnamon
½ tsp. grated lemon peel

Cream butter and sugar in large bowl; beat in egg yolks. Gradually stir in flour until dough almost cleans side of bowl. Shape dough into ball. Refrigerate wrapped in plastic wrap until cold, about 1 hour. Heat oven to 350°. Roll dough on lightly floured surface ⅛-inch thick. Cut out cookies with 2-inch cutter. Place on greased cookie sheets. Beat egg whites in small bowl until foamy. Gradually beat in 1¼ cups sugar; beat until stiff and glossy. Fold in almonds, cinnamon and lemon peel. Spoon mixture into pastry bag fitted with ½-inch star tip; pipe a rosette onto center of each cookie. Bake until tops are light brown, 12 to 14 mintues. Cool.

Don't worry when you stumble. Remember, the worm is about the only thing that can't fall.

SUGAR-FREE PINEAPPLE COOKIES

1 c. margarine
Brown sugar substitute to equal 1 cup (see package label)
Sugar substitute to equal 1 cup (see package label)
2 eggs
2 c. all-purpose flour
1¾ c. stone ground whole wheat flour
2 tsp. baking powder
½ tsp. salt
1 c. crushed pineapple in it's own juice (not syrup)
1 tsp. vanilla
1 c. pecan pieces
Icing

Cream margarine and sugar substitutes. Beat in eggs. Stir in dry ingredients. Drain pineapple, reserving juice. Stir in pineapple and vanilla. Add a little pineapple juice, if needed, for drop-cookie consistency. Stir in nuts. Drop by teaspoon onto greased cookie sheet. Bake in 400° oven for 10 minutes, or until lightly browned. Cool before icing. For icing, stir powdered sugar substitute and crushed pineapple together to consistency for light icing. Makes 4 to 5 dozen cookies. Each unfrosted cookie has about 60 calories, frosted cookies have about 80 calories.

WHOLE WHEAT CHOCOLATE CHIP COOKIES

1½ c. brown sugar
1 c. butter, softened
1 tsp. vanilla
3 eggs
2½ c. stone ground whole wheat flour
⅓ c. instant dry milk
1 tsp. soda
½ tsp. salt
2 c. semi-sweet chocolate pieces
¾ c. chopped pecans
½ c. sunflower nuts

Heat oven to 350°. In large bowl, blend first four ingredients until smooth. Stir in flour, dry milk, soda and salt. Stir in remaining ingredients. Drop by teaspoonfuls, two inches apart onto greased cookie sheets. Bake 10 to 14 minutes until light brown. Makes 5 to 6 dozen cookies.

WHOLE WHEAT HONEY COOKIES

4 tblsp. butter
½ c. honey
1½ c. stone ground whole wheat flour
1 c. dry milk
½ tsp. baking soda
½ tsp. salt
2 tblsp. water
2 eggs
½ tsp. vanilla
1 c. chocolate chips
½ c. sunflower seeds
¼ c. chopped peanuts

Combine ingredients in order given. Drop by teaspoon onto lightly greased cookie sheet. Bake in 350° oven for 12 minutes. Makes 3 to 4 dozen cookies.

WHOLE WHEAT COOKIES

1 c. all-purpose flour
1 c. stone ground whole wheat flour
1½ tsp. baking powder
½ tsp. salt
1 c. brown sugar
⅔ c. shortening
1 egg
3 tblsp. milk
1 tblsp. grated orange peel

Stir together flour, baking powder and salt. Cream sugar and shortening until light and fluffy. Beat in egg, milk and orange peel. Thoroughly blend in flour mixture. Drop by teaspoonfuls, 2 inches apart, onto greased baking sheets. Bake in preheated 375° oven 8 to 10 minutes, or until golden brown. Remove immediately from baking sheet and cool on wire rack. Makes about 4 dozen cookies.

WHOLE WHEAT OATMEAL COOKIES

1 c. raisins
1 c. shortening
¾ c. sugar
¾ c. packed brown sugar
2 eggs
1½ c. stone ground whole wheat flour
1 tsp. baking soda
1 tsp. salt
½ tsp. ground nutmeg
½ tsp. cinnamon
2 c. quick cooking rolled oats
½ c. chopped walnuts

In small saucepan cover raisins with water. Bring to boiling; remove from heat. Cover and let stand 5 minutes. Put through fine blade of food grinder. Set aside. In mixer bowl cream shortening and sugars until light and fluffy. Beat in eggs. Stir together whole wheat flour, baking soda, salt, nutmeg and cinnamon. Add to creamed mixture. Stir in oats, walnuts and ground raisins. Drop by teaspoonfuls onto greased cookie sheet. Bake at 350° for 10 to 12 minutes. Makes 7 to 8 dozen cookies.

WHOLE WHEAT SUGAR COOKIES

1 c. sugar
½ c. margarine
1 egg
2 c. stone ground whole wheat flour
1 tsp. baking powder
½ tsp. baking soda
½ tsp. salt
½ tsp. nutmeg
1 tblsp. grated orange peel
2 tblsp. milk
1 tsp. vanilla
½ tsp. cinnamon and 2 tsp. sugar mixed for topping

Cream sugar and shortening; then add egg. Mix dry ingredients together on sheet of wax paper. Add to creamed mixture with liquid ingredients. Blend well. Shape in 1-inch balls and place on lightly greased cookie sheet 2 inches apart. Flatten slightly with glass dipped in cinnamon-sugar mixture. Bake 8 to 10 minutes in preheated oven or until golden brown.

WHOLE WHEAT SUGAR COOKIES

1½ c. firmly packed brown sugar
½ c. margarine, softened
½ c. shortening
1 tsp. vanilla
2 eggs
2½ c. stone ground whole wheat flour
½ c. wheat germ
2 tsp. baking powder
½ tsp. salt
½ c. sugar
1 tsp. cinnamon

In large bowl, combine brown sugar, margarine, shortening, vanilla and eggs; blend well. Lightly spoon flour into measuring cup; level off. Stir in flour, wheat germ, baking powder and salt. For easier handling, refrigerate dough about 30 minutes. Heat oven to 400°. Shape dough into 1 inch balls. Combine sugar and cinnamon; roll balls in sugar mixture. Place 2 inches apart on ungreased cookie sheet. Bake at 400° for 7 to 10 minutes or until light golden brown. Cool 1 minute before removing from cookie sheet. Makes 6 dozen cookies.

WORLD'S GREATEST COOKIES

1 c. butter OR margarine
1 c. peanut butter
1 c. white sugar
1 c. brown sugar (firmly packed)
2 eggs
2 c. stone ground whole wheat flour
1 tsp. soda
1 6-oz. pkg. chocolate chips
¼ c. cornstarch

Cream butter or margarine and peanut butter together. Gradually add the white and brown sugars and cream until blended. Add eggs, one at a time and beat until smooth. Stir flour with cornstarch and soda. Add to the mixture. Stir in chocolate chips. Drop from a teaspoon onto a greased baking sheet; then slightly flatten cookie dough with the back of spoon. Bake in moderate oven 325° for about 15 minutes. Makes 6 dozen cookies 2 inches in diameter.

ZUCCHINI SPICE COOKIES

½ c. shortening
¾ tsp. ground cinnamon
¼ tsp. nutmeg
¼ tsp. ground cloves
1 c. packed dark brown sugar
1 egg
¾ c. all-purpose flour
1 c. stone ground whole wheat flour
2 tsp. baking powder
½ tsp. salt
¼ c. milk
1½ c. shredded unpared zucchini
½ c. chopped nuts

In a large mixing bowl, cream shortening, spices and brown sugar until light and fluffy. Beat in egg. Mix flour, baking powder and salt. Add alternately with milk to creamed mixture. Stir in zucchini and nuts. Drop by tablespoonfuls onto greased baking sheet. Bake in 375° oven for 12 to 15 minutes or until golden. Remove from baking sheet and cool on a wire rack. Store in a tightly covered container. Makes 4 dozen.

ZUCCHINI / OATMEAL COOKIES

¾ c. butter OR margarine
¾ c. brown sugar
1 egg
1 tsp. vanilla
½ tsp. baking soda
½ tsp. salt
1 tsp. cinnamon
1½ c. rolled oats
¾ c. grated zucchini
1½ c. stone ground whole wheat flour
1 c. raisins

Cream together butter or margarine, brown sugar, egg, vanilla, baking soda, salt and cinnamon until smooth and creamy. Stir in oats, zucchini, flour and raisins until well combined. Drop by tablespoonfuls onto a lightly greased or non-stick cookie sheet. Bake at 350° 10 minutes or until lightly browned and set. Cool on a wire rack.

APPLE MOLASSES BARS

1 c. butter
½ c. brown sugar
1 c. white sugar
½ c. molasses
2 eggs
1 c. applesauce
1½ c. stone ground whole wheat flour
1½ c. all-purpose flour
2 tsp. baking powder
½ tsp. soda
½ tsp. salt
½ tsp. cinnamon
½ c. dairy sour cream

Frosting:
1 3-oz. cream cheese
¼ c. butter
2 c. confectioner's sugar
1 tblsp. milk
1 tsp. vanilla

Cream butter, sugars and molasses. Add eggs, one at a time, beating until mixture is fluffy. Stir in applesauce. Combine flour, baking powder, soda, salt and cinnamon. Add alternately with sour cream to creamed mixture. Spread in greased, large 12 x 17 x 1-inch jelly roll pan or two 8 x 10-inch pans. Bake at 350° for 25 to 30 minutes. Frost with cream cheese frosting.

APPLE NUT BARS

3 medium apples, peeled, seeded, and ground or chopped coarsely
½ c. water
2 eggs
1 c. granulated sugar
½ c. shortening
⅔ c. stone ground whole wheat flour
1 c. all-purpose flour
½ tsp. baking soda
1 tsp. baking powder
1 tsp. cinnamon
½ c. chopped nuts

Put apples and water into saucepan and cook over medium heat for five minutes. Set aside to cool. Combine eggs, sugar, and shortening, and add cooled apples. Sift together dry ingredients; add to apple mixture and mix well. Add chopped nuts. Bake in a greased 9 x 13-inch pan for 25 to 30 minutes at 350°, until a toothpick comes out clean. Cool slightly and frost with a thin confectioner's sugar icing. Cut in squares and serve.

APRICOT BARS

¾ c. stone ground whole wheat flour
¾ c. all-purpose flour
1 tsp. baking powder
1½ c. quick cooking oats
¼ tsp. salt
1 c. brown sugar
¾ c. butter OR margarine
¾ c. apricot preserves

Stir together dry ingredients. Stir in oats and sugar. Add margarine and crumble. (Mix together until pieces are very small.) Put two-thirds of the mixture into a 9 x 9-inch pan. Pat down so it makes crust on pan. Spread top with apricot preserves and sprinkle with rest of oats-sugar mixture. Bake at 350° for 30 to 35 minutes. Let cool before cutting.

APPLE SQUARES

1 c. all-purpose flour
1 c. stone ground whole wheat flour
2 c. brown sugar
½ c. softened butter
1 c. chopped nuts
1½ tsp. cinnamon
1½ tsp. soda
½ tsp. salt
1 c. dairy sour cream
1 tsp. vanilla
1 egg
2 c. chopped apples

Combine flour, sugar and butter. Mix until crumbly, add nuts. Press 2¾ cups of this mixture into a 9 x 13-inch pan. To remaining crumb mixture add cinnamon, soda, salt, sour cream, vanilla and egg; blend well. Stir in peeled and finely chopped apples. Spoon evenly over crust. Sprinkle with rest of crumb mixture. Bake at 350° for 25 to 30 minutes until toothpick inserted in center comes out clean.

BARS OF IRON

⅓ c. margarine
½ c. **each** sugar, light molasses
1 egg
1 ¼ c. stone ground whole wheat flour
¼ c. nonfat dry milk
¼ c. toasted wheat germ
1½ tsp. baking powder
½ tsp. **each** soda, salt, ginger
½ c. liquid milk
1 c. rolled oats
1 c. dark raisins, chopped
½ c. golden raisins, chopped
1 c. sliced almonds

Stir together butter, sugar, molasses and egg. Combine flour, dry milk, wheat germ, baking powder, soda, salt and ginger. Mix lightly. Blend into creamed mixture alternately with liquid milk. Stir in oats, raisins and half the almonds. Turn into a greased 9 x 13-inch baking pan, spread evenly. Sprinkle with remaining almonds. Bake at 350° for 30 minutes, until bars test done. Cool in pan. Cut into 4 x 1-inch bars. Makes 3 dozen.

BLONDE BROWNIES

1 c. margarine
3 c. brown sugar
3 eggs, beaten
3 tsp. vanilla
1½ tsp. salt
1½ tsp. baking powder
½ tsp. soda
2 c. all-purpose flour
1 c. stone ground whole wheat flour

Melt margarine in saucepan, add brown sugar, then add eggs. Mix in the dry ingredients. Add nuts, chocolate chips or coconut as desired. Bake in large sheet cake pan at 350° for 25 minutes or until center falls. Do not overbake.

BUTTERSCOTCH BANANA BARS

1 c. granulated OR brown sugar
¾ c. shortening
1 egg
¾ c. all-purpose flour
½ c. stone ground whole wheat flour
½ tsp. salt
½ tsp. baking soda
1 pkg. (6 oz.) butterscotch chips
1 c. mashed banana
1 c. cut dates, optional
¾ c. chopped nuts

Preheat oven to 375°. Cream sugar and shortening; beat in egg. Combine remaining ingredients, except bananas. Add alternately with bananas and spread into greased 10 x 15-inch pan. Bake 20 to 25 minutes until bars test done. Makes about 60 small bars.

CALCIUM BARS

1 c. shortening
2½ c. brown sugar
4 eggs
2 tsp. maple flavoring
½ c. all-purpose flour
1 c. stone ground whole wheat flour
1 tsp. baking powder
½ tsp. salt
½ tsp. soda
2 c. dry skim milk
1 c. chopped walnuts

Cream together shortening and sugar. Add eggs and flavoring. Beat in flour, etc. Add milk. Bake in 11 x 15-inch pan approximately 30 minutes. Cut while warm and roll in powdered sugar. Good frozen.

CARROT ORANGE BARS

¾ c. shortening
¾ c. sugar
1 c. (¾ lb.) cooked mashed carrots
1 egg
1 c. unsifted all-purpose flour
1 tsp. baking powder
2 c. stone ground whole wheat flour
¾ tsp. salt
1½ tsp. grated orange peel
¾ c. flaked coconut
⅓ c. orange juice
GLAZE:
3 tblsp. orange juice
¾ tsp. grated orange peel
¾ c. confectioner's sugar

Preheat oven to 350°. Grease a 9 x 13-inch baking pan; set aside. In large mixer bowl with electric mixer at medium speed, cream shortening and sugar. Add carrots and egg and mix until well blended. In a separate bowl combine flour, baking powder and salt. Gradually add to creamed mixture until combined. Add orange peel, coconut and orange juice; mix until well blended. Spread batter evenly into a prepared pan. Bake 20 to 25 minutes or until toothpick inserted in the center comes out clean. Cool completely, spread with glaze. Cut into 3 x 1-inch bars. Makes 3 dozen bars. Glaze: In a small bowl combine orange juice, orange peel and confectioner's sugar until smooth. Freezes well.

CARROT-YOGURT SQUARES

1 c. honey
¾ c. oil
8-oz. carton plain yogurt
2 eggs
1 c. all-purpose flour
1 c. stone ground whole wheat flour
2 tsp. baking powder
¼ tsp. salt
1½ tsp. cinnamon
1 c. grated carrots
1 c. chopped nuts
1 c. raisins

Frosting:

3-oz. pkg. cream cheese, softened
3 c. powdered sugar
¼ tsp. salt
2-3 tblsp. milk
1 tsp. vanilla

Heat oven to 350°. Grease 9 x 13-inch pan. In large bowl, combine honey, oil, yogurt and eggs; beat well. Lightly spoon flour into measuring cup; level off. Add all-purpose flour, whole wheat flour, baking powder, ¼ tsp. salt and cinnamon to liquid mixture; blend well. Stir in carrots, nuts and raisins; mixing well. Pour into prepared pan. Bake at 350° for 30-40 minutes or until toothpick inserted in center comes out clean. Cool. Meanwhile, in small bowl, beat cream cheese. Beat in powdered sugar, ¼ tsp. salt, milk and vanilla until frosting is smooth and creamy. If necessary, add additional milk until of spreading consistency. Spread over bars. Makes 36 bars.

CHERRY COCONUT BARS

1 c. butter
¼ c. sugar
1 c. all-purpose flour
1 c. stone ground whole wheat flour
4 eggs, beaten
2 c. sugar
1 c. coconut
½ c. all-purpose flour
1 tsp. baking powder
½ jar maraschino cherries (10 oz.) drained and sliced
1 tsp. vanilla
1 c. chopped nuts

Mix butter, ¼ c. sugar and flours together. Pat into bottom of 9 x 13-inch pan. Add sugar and dry ingredients to beaten eggs. Add cherries, nuts and vanilla. Mix well and spread over crust. Bake 20 to 30 minutes at 325°. Frost. May use juice from cherries in frosting.

CINNAMON COFFEE BARS

¼ c. soft shortening
1 c. brown sugar
1 egg, beaten
½ c. coffee
¾ c. all-purpose flour
¾ c. stone ground whole wheat flour
1 tsp. baking powder
¼ tsp. baking soda
¼ tsp. salt
½ tsp. cinnamon
½ c. raisins
¼ c. chopped nuts
Confectioner's sugar
Milk

Preheat oven to 350°. Cream shortening, brown sugar and egg. Stir in coffee. Sift flour with baking powder, baking soda, salt and cinnamon. Add to brown sugar mixture and mix well. Blend in raisins and nuts. Grease and flour a 9 x 13-inch pan. Bake 15 to 20 minutes. Frost while warm with a creamy icing; blend confectioner's sugar with cream or milk to a spreading consistency.

CRANBERRY OATMEAL BARS

¾ c. stone ground whole wheat flour
¾ c. sifted all-purpose flour
⅛ tsp. salt
1 c. packed brown sugar
1 can (1 lb.) whole berry cranberry sauce
½ c. crushed pineapple, drained
½ tsp. baking soda
1½ c. rolled oats (quick or regular)
¾ c. butter OR margarine
¼ tsp. vanilla
½ c. chopped pecans (optional)

Sift flour, soda and salt together. Combine with oats and sugar. Using pastry blender or two knives, cut in butter or margarine until mixture is crumbly. Press half of the mixture firmly in bottom of a greased 7 x 11 x 1½-inch pan. Combine cranberry sauce, pineapple, nuts and vanilla. Pour over crust. Spread evenly. Sprinkle with remaining mixture. Bake in 400° oven until light brown, about 25 minutes. Let cool, cut in bars. Makes 30 bars.

DATE BARS

⅓ c. butter OR margarine at room temp.
½ c. plus 2 tblsp. honey
1¼ c. stone ground whole wheat flour
1¼ c. rolled oats
½ tsp. salt
½ c. chopped nuts
1¾ c. chopped dates
¾ c. water

Cream butter, then cream in honey. When smooth, add dry ingredients. Press half this mixture into a buttered 8-inch square pan. To make filling, combine dates and water in a saucepan and simmer gently 5-10 minutes, until water is absorbed. Spread filling over oat mixture, then crumble remaining oat mixture over filling. Bake 30 minutes at 325° until golden. Cool before serving.

FRESH ZUCCHINI-LEMON BARS

3 eggs
1¼ c. sugar
1 c. vegetable oil
¼ c. freshly squeezed lemon juice
2 tsp. grated lemon rind
1 c. coarsely chopped walnuts
1¾ c. finely shredded, unpeeled zucchini (about 3 medium)
1 c. unsifted flour
1 c. stone ground whole wheat flour
2 tsp. baking soda
1 tsp. salt
1 tsp. baking powder

In medium bowl combine eggs, sugar, oil, lemon juice and rind; beat by hand until well blended. Add zucchini; mix well. Sift together flour, baking soda, salt and baking powder; stir into zucchini mixture. Fold in nuts. Pour into greased 9 x 13-inch baking pan. Bake in a 350° oven, 45 to 50 minutes or until it springs back when pressed lightly with tip of finger. Cool in pan. Cut in bars to serve.

GERMAN CHOCOLATE BARS

2¼ c. stone ground whole wheat flour
¼ c. cornstarch
⅓ c. powdered milk
1 tsp. baking soda
½ c. water, boiling
1 4-oz. pkg. German sweet chocolate
2 c. sugar
9 egg yolks
1 c. butter
1 tblsp. vinegar
½ c. water

Coconut Frosting:
1 c. sugar
3 egg yolks
¾ c. evaporated milk
¾ c. butter
1 7-oz. pkg. coconut
½ c. nuts

Sift whole wheat flour, cornstarch, powdered milk and soda. Break German chocolate bar into pieces and soften in top of double boiler. Combine sugar, egg yolks and butter. Add vinegar, water and chocolate. Stir in flour mixture and blend well. Bake in lightly greased pan for 40 minutes. This recipe may also be made into a layer cake by baking in two 9-inch round cake pans. Coconut Frosting: Blend sugar, egg yolks and milk in small pan. Add butter. Bring to boil, stirring constantly. Add coconut and nuts. Frost bars.

GINGERSNAP BARS

¾ c. shortening
1 c. all-purpose flour
1 c. stone ground whole wheat flour
1 c. sugar
2 tsp. soda
1 tsp. cinnamon
½ tsp. cloves
½ tsp. ginger
½ tsp. salt
¼ c. molasses
1 egg
2 tblsp. sugar (for topping)

Heat oven to 375°. Grease 10 x 15-inch jelly roll pan or cake pan. In a large saucepan, melt shortening; cool 5 minutes. Add remaining ingredients except the 2 tblsp. of sugar; mix well. Press in bottom of prepared pan. Sprinkle with remaining sugar. Bake for about 10 minutes. Do not overbake. Let stand 5 minutes, and then cut into bars. Allow to cool completely before serving. Makes 48 bars.

GRANOLA BARS

6 tblsp. butter OR margarine
½ c. packed brown sugar
¼ c. honey
1 egg
1 tsp. vanilla
½ c. all-purpose flour
½ c. stone ground whole wheat flour
¾ tsp. baking powder
½ tsp. salt
½ tsp. baking soda
1½ c. bite-sized crispy bran squares crushed to ½ cup
½ c. raisins
½ c. flaked coconut
¼ c. and 3 tblsp. sunflower nuts
2 tblsp. sesame seeds

Preheat oven to 350°. Grease 9 x 13 x 2-inch baking pan. Cream together butter and sugar. Beat in honey, egg and vanilla. Stir together flour, baking powder, salt and baking soda and cereal. Add to creamed mixture. Stir in raisins, coconut and ¼ c. sunflower nuts. Spread evenly in pan. Sprinkle sesame seeds and remaining 3 tblsp. sunflower nuts over the top. Bake 15 to 20 minutes or until toothpick inserted in center comes out clean and top is lightly browned. Cool. Cut into bars. Makes 24 bars.

GRANOLA BARS

½ c. stone ground whole wheat flour
3 c. toasted quick oats
1 c. raisins (optional)
1 c. chopped nuts (optional)
⅔ c. margarine, melted
¼ c. brown sugar, firmly packed
⅓ c. honey, corn syrup, OR molasses
1 egg, beaten
½ tsp. vanilla
½ tsp. salt

First toast oats by placing on a cookie sheet. Bake in 350° oven 15 - 20 minutes or until light golden brown. Combine all ingredients and mix well. Place firmly into well greased 15 x 10-inch jelly roll pan. Bake in 350° oven 20 minutes. Cool. Cut into bars. Store in tightly covered container in cool dry place or in refrigerator.
Variations: Add ½ c. dry coconut, flaked or shredded. Substitute ½ c. carob chips for ½ c. raisins. Substitute ½ c. sunflower seeds for ½ c. nuts. Makes 40 bars 2-inch x 2-inch or 80 bars 1-inch x-2 inch.

OATMEAL BARS

¾ c. shortening
1 c. firmly packed brown sugar
½ c. granulated sugar
¼ c. water
1 tsp. vanilla
½ tsp. rum flavoring (optional)
½ c. all-purpose flour
½ c. stone ground whole wheat flour
1 tsp. salt
½ tsp. soda
3 c. uncooked oats
½ c. chocolate chips
½ c. butterscotch chips

In a large bowl combine shortening, brown sugar, sugar, water, vanilla and rum flavoring. Stir until fluffy. Add flour, salt and soda, mixing well. Stir in oats and chips. Bake in two 8 x 8-inch well-greased pans at 350° for about 20 minutes.

PUMPKIN BARS

4 eggs, beaten
1 16-oz. can pumpkin
1⅔ c. sugar
1 c. cooking oil
1 c. all-purpose flour
1 c. stone ground whole wheat flour
2½ tsp. baking powder
2 tsp. cinnamon
1 tsp. soda
1 tsp. salt

Combine eggs, pumpkin, sugar and oil; beat until light and fluffy. Sift together the dry ingredients and add to egg mixture. Bake in 11 x 17 x 1-inch ungreased pan at 350° for 20 - 25 minutes. Frost with cream cheese frosting.

SNACK-ATTACK SQUARES

¾ c. butter OR margarine
1½ c. firmly packed light brown sugar
2 eggs
1 tsp. vanilla
1 c. stone ground whole wheat flour
1 c. all-purpose flour
1½ c. raisin bran cereal
½ tsp. baking soda
½ tsp. baking powder
½ tsp. salt
1 c. chopped nuts
1 c. multicolored plain chocolate candies

Blend together butter and sugar until light and fluffy; blend in eggs and vanilla. Add combined flours, cereal, baking soda, baking powder and salt; mix well. Stir in ¾ c. nuts and ⅔ c. candies. Spread dough into greased 9 x 13-inch pan. Sprinkle with remaining ¼ c. nuts and ⅓ c. candies; press in lightly. Bake at 350° for 30 to 35 minutes or until edges are lightly browned. Cool thorougly; cut into squares.

SPICED RAISIN SQUARES

2 c. stone ground whole wheat flour
1 c. raisins
½ c. chopped nuts
1½ tsp. soda
½ tsp. ginger
¼ tsp. nutmeg
¼ tsp. allspice
2 eggs
¼ c. butter, melted
3 tblsp. honey
1 c. buttermilk or sour milk
1 tsp. finely grated orange peel

Combine first seven ingredients. Combine rest of ingredients. Mix with dry ingredients. Place in greased 8 x 8-inch pan. Bake at 350° for 30 to 35 minutes. Cut in squares.

SPLIT LEVELS

1 c. (6-oz.) semi-sweet chocolate bits
1 pkg. (3-oz.) cream cheese
1/3 c. evaporated milk or light cream
1/2 c. chopped walnuts
1/4 tsp. almond extract
1 1/4 c. stone ground whole wheat flour
1/2 tsp. baking powder
1/4 tsp. salt
3/4 c. sugar
1/2 c. butter OR margarine, softened

Combine chocolate bits, cream cheese and milk in saucepan. Melt over low heat, stirring constantly. Remove from heat and stir in nuts and almond extract. Set aside. Combine remaining ingredients; blend at low speed just until particles are fine. Press half of mixture in greased 9 x 9-inch pan. Spread with chocolate mixture. Sprinkle with remaining crumbs. Bake at 375° for 20 - 25 minutes. Cool; cut in bars.

WHEAT 'N FRUIT BARS

2 c. rolled oats (quick OR old fashioned, uncooked)
1/2 c. all-purpose flour
1/2 c. stone ground whole wheat flour
3/4 c. margarine OR butter, melted
1/3 c. firmly packed brown sugar
1/3 c. chopped nuts
1/2 tsp. salt (optional)
1/2 tsp. baking soda
1 10 to 12-oz. jar favorite fruit preserves

Heat oven to 350°. Grease 11 x 7 inch glass baking dish. Combine all ingredients except preserves; mix well. Reserve 3/4 c. oats mixture; press remaining mixture onto bottom of prepared dish. Bake 10 minutes. Spread preserves evenly over partially baked base to within 1/2 inch of edge of pan; sprinkle with reserved oats mixture. Bake 20 to 22 minutes or until golden brown. Cool; cut into bars. Store tightly covered at room temperature.

WHOLE WHEAT BANANA BARS (or cookies)

1 c. shortening, part butter
1 c. sugar
2 eggs
1 c. mashed ripe bananas
1/2 c. buttermilk
1 tsp. vanilla
3 c. stone ground whole wheat flour (OR 2 c. whole wheat & 1 c. all-purpose)
1 1/2 tsp. soda
1/2 tsp. salt

Glaze:
1/4 c. boiling water
1 c. confectioner's sugar
1 tsp. vanilla

Cream together shortening, sugar and eggs. Mix in bananas, buttermilk and vanilla. Mix together flour, soda and salt and stir into banana mixture. Batter will be thick. Spread batter in greased 10 x 15-inch jelly roll pan. Bake in 375° oven 20-25 minutes or until cake springs back when lightly touched. Combine glaze ingredients and pour over hot cake. (Chopped nuts may be sprinkled over cake before baking.)
For cookies: Chill dough at least 1 hour. Drop by teaspoonfuls onto greased cookie sheet. Bake 8 - 10 minutes in 375° oven. Remove from cookie sheets and place on racks over waxed paper. Drizzle with glaze. If desired, add chopped nuts before chilling dough.

WHOLE WHEAT JAM SQUARES

2 c. uncooked oatmeal
1 c. all-purpose flour
¾ c. stone ground whole wheat flour
1 c. butter OR margarine
1 c. brown sugar
½ c. chopped nuts
1 tsp. cinnamon
¾ tsp. salt
½ tsp. soda
¾ c. preserves, jam OR jelly

Combine all ingredients except preserves in large mixing bowl; beat at low speed on electric mixer until mixture is crumbly. Reserve 2 cups mixture; press remaining onto bottom of greased 9x 13-inch baking pan. Spread preserves, jam or jelly evenly over base; sprinkle with reserved mixture. Bake in preheated 400° oven 25 to 30 minutes. Cool. Cut into squares. Makes about 2 dozen squares.

ZUCCHINI BARS

3 eggs
⅔ c. oil
1½ c. sugar
2 tsp. vanilla
1¼ c. stone ground whole wheat flour
1 c. all-purpose flour
2 tsp. soda
1 tsp. salt
½ tsp. baking powder
2 c. shredded zucchini (drained)
¾ c. raisins AND/OR dates
½ c. chopped nuts

Combine eggs, oil, sugar and vanilla. Combine flour, soda, salt and baking powder. Add to liquid ingredients. Add zucchini, raisins, dates and nuts. Bake in 15 x 10 x 1-inch pan at 350° for 30 minutes. Cool and frost.

Frosting:
3-oz. cream cheese (soft)
2½ c. confectioner's sugar
1 tsp. vanilla
1 tblsp. (or more) milk

Combine all ingredients and spread on bars.

DESSERTS

APPLE BROWN BETTY

- 4 c. apples, sliced
- 1 tblsp. lemon juice
- 3/4 c. brown sugar
- 1 tsp. cinnamon
- 1/4 tsp. nutmeg
- 1/4 tsp. salt
- 1 tsp. grated lemon peel
- 2 c. stale whole wheat bread crumbs
- 1/2 c. hot water
- 3 tblsp. melted butter

Sprinkle apples with lemon juice. Mix sugar, spices, salt and lemon peel together. Put half of the apples in bottom of a buttered 1 1/2 qt. casserole or 8 x 8 x 2-inch baking dish. Cover with half of the bread crumbs; then half the sugar mixture. Repeat layers. Pour hot water and melted butter over top. (Use more water if bread is quite dry.) Bake covered at 350° for 40 minutes or until apples are nearly tender. Remove cover and bake 10 to 15 minutes longer to brown the top. Serve warm with cream or ice cream.

APPLE CRISP

- 6 cooking apples
- 1 c. sugar
- 1/2 c. butter OR margarine
- 1 1/2 tsp. cinnamon
- 1 c. stone ground whole wheat flour

Peel and slice apples into 9 x 9 inch baking dish or 9 inch deep dish pie plate. Combine sugar, butter, cinnamon and whole wheat flour; blend thoroughly. Sprinkle over sliced apples. Bake 45 minutes in 350° oven. Makes 6 to 8 servings.

APPLE CRISP

- 8 apples
- Juice of 1 lemon
- 1 tsp. cinnamon
- 2 tblsp. stone ground whole wheat flour
- 3/4 c. raisins
- Apple Juice

Topping:
- 1 c. rolled oats
- 1/3 c. raw wheat germ
- 1/2 c. stone ground whole wheat flour
- 1/2 tsp. salt
- 2 tsp. cinnamon
- 1/3 c. brown sugar
- 3 tblsp. each, soft butter and oil

Preheat oven to 375°. Slice apples until you have enough to fill a greased 9 x 13-inch baking dish. Mix the apples in a bowl with lemon juice, cinnamon, flour and raisins. Return them to the baking dish, adding enough apple juice to cover the bottom. Mix oats, wheat germ, flour, salt, cinnamon, sugar, butter and oil in a bowl and press onto top of apples. Bake for 25 minutes, or until apples are soft. Makes 8 servings.

APPLE ROLL-UPS

1 c. stone ground whole wheat flour
1 c. all-purpose flour
2 tsp. baking powder
1 tsp. salt
2/3 c. shortening
1/2 c. milk
2 1/2 c. grated raw apples

Syrup:
2 c. sugar
2 c. water
1/4 tsp. cinnamon
1/4 tsp. nutmeg
6 to 10 drops red food coloring
2 tblsp. margarine

Sift dry ingredients; cut in shortening till mixture resembles coarse crumbs. Add milk all at once and stir until flour is moistened. On lightly floured surface roll about 1/4thick into 18 x 12-inch rectangle. Spread on grated raw apples and roll like jelly roll. Cut into 8 or 10 pieces and place in plastic bags and freeze. When ready to bake, place in ungreased pan.
Syrup: Make syrup by bringing all ingredients to boiling point and pour over the frozen apple rolls and bake 35 to 40 minutes at 375°. Serve with cream or milk.

BAKED WHEAT PUDDING

1 qt. milk
5 tblsp. cracked wheat
1 tsp. salt
1/2 tsp. ginger
1/2 tsp. nutmeg
2 tblsp. butter
1 c. molasses, dark
2 eggs
1/2 c. seeded raisins

Scald milk. Pour over cracked wheat gradually, stirring constantly till smooth and creamy. Cook 20 minutes. Pour over well beaten eggs. Stir well, add other ingredients, and pour into a greased baking dish. Bake 1 1/2 hours at 325°. Serve with vanilla ice cream or whipped cream.

BEIGE CREAM PUFFS

1 c. water
1/2 c. butter OR margarine
1 c. stone ground whole wheat flour
4 eggs

Boil water with margarine and stir until dissolved. Turn heat off and add flour, stirring briskly until dough sticks to itself. Add eggs, one at a time, beating well after each egg is added. Place balls of dough three inches apart on lightly greased cookie sheets. Bake at 450° for 15 minutes, then reduce temperature to 325° and bake for 25 minutes more. Remove puffs, turn oven off, split or cut with sharp knife and return to oven with door ajar for about 1/2 hour to dry. Fill with favorite filling.

BREAD PUDDING WITH VANILLA SAUCE

4 c. toasted whole wheat bread crumbs
4 c. warm milk
1 c. sugar
¼ tsp. nutmeg
¼ tsp. salt
¼ c. coconut
¼ c. raisins
4 oz. soft butter OR margarine
1 tsp. vanilla
9 eggs

Combine all of the above ingredients except eggs and mix ingredients well. Beat eggs separately until light and beat into mixture. Pour into a 4-quart buttered casserole and place in a pan of warm water in a 350° oven for about 1 hour, or until golden brown and pudding is set. Serve warm with Vanilla Sauce: Whip together until light and fluffy, 9 oz. soft vanilla ice cream, 9 oz. Cool Whip and ½ tsp. vanilla. Serves 8 to 10.

BROWNIE PUDDING CAKE

1½ c. stone ground whole wheat flour
1 c. sugar
3 tblsp. cocoa
3 tsp. baking powder
¼ tsp. salt
¾ c. milk
1½ tblsp. melted butter
2 tsp. vanilla
¾ c. chopped walnuts
5 drops red food color
½ c. brown sugar, firmly packed
¼ c. sugar
3 tblsp. cocoa
1¾ c. boiling water
Vanilla ice cream

Sift together flour, 1 c. sugar, 3 tblsp. cocoa, baking powder and salt into a bowl. Add milk, butter and vanilla; beat well. Stir in walnuts and food color. Spread batter in greased 13 x 9 x 2-inch baking pan. Combine brown sugar, ¼ c. sugar and 3 tblsp. cocoa. Sprinkle over batter. Pour boiling water over all. Bake at 350° for 40 minutes or until cake tests done. Serve warm topped with vanilla ice cream. Makes 12 servings.

CHERRY PIZZA DESSERT

Crust:
1 pkg. dry yeast
1 tblsp. water
1 c. butter OR margarine
1 c. stone ground whole wheat flour
¾ c. all-purpose flour
2 tblsp. cream or half-and-half

Filling:
8 oz. cream cheese
½ c. sugar
2 eggs
1 tsp. vanilla

Topping:
1 can cherry or pineapple pie filling
Sour cream

Dissolve yeast in water (you may need to add a bit more). Cream butter, beat in yeast mixture, and gradually work in flour and cream. Press into a 9 x 12-inch pan and let stand 20 minutes. Bake in a preheated 350° oven for 30 minutes. Meanwhile, cream the cheese and sugar together; add eggs and vanilla. Spread this in a thin layer on top of the baked crust. Raise oven temperature to 375° and bake 10 minutes longer or until filling is set. Cool and cover with pie filling. Serve with sour cream.

CROCKPOT WHEAT DESSERT

1 c. uncooked wheat kernels
½ c. undrained crushed pineapple
Water
1 c. miniature marshmallows
¼ c. chopped pecans
1 c. prepared topping or whipped cream

Place wheat in a crockpot and cover with water. Cook for two hours; stir and add 1 cup hot water. Stir well. Some wheat is drier and will take more water. Cool. Fold in remaining ingredients and serve.

DATE PUDDING

Syrup:
1 c. brown sugar
2½ c. water
2 tblsp. butter

Batter:
½ c. brown sugar
⅞ c. stone ground whole wheat flour
2 tsp. baking powder
¼ tsp. salt
½ c. milk
1 c. chopped dates
½ c. chopped nuts

Cook brown sugar and water for syrup together for 3 minutes. Then add butter. Pour syrup into bottom of baking dish. Set aside. Sift dry ingredients for batter together, add milk, then fold in nuts and dates. Pour batter on top of syrup in baking dish and bake at 350° for 35 to 40 minutes. Serve plain or with whipped cream.

Genius is not inspiration as much as perspiration.

DREAMY CREAM PUFFS

1 c. water
¾ to 1 stick butter
1 tsp. salt
½ c. stone ground whole wheat flour
½ c. all-purpose flour
Pinch nutmeg (optional)
4 large eggs (room temperature)

Place water in 1½-quart heavy bottomed saucepan. Add butter, cut in pieces, salt and nutmeg. Bring to boil and simmer until butter has melted. Remove from heat and immediately pour in all the flour. Beat vigorously with wooden spoon for several seconds to blend thoroughly. Return to medium-hot burner and beat 1 to 2 minutes. Remove saucepan from heat and make a well in the center of the paste with the spoon. Immediately break an egg into the center of the well. Beat it into the paste for several seconds until it has absorbed. Continue with the rest of the eggs, beating them in one by one; the third and fourth eggs will be absorbed more slowly. Continue beating until mixture is smooth and satiny. Drop from spoon or force through pastry tube onto greased cookie sheets. Bake in 425° oven about 20 minutes. For large puffs, reduce heat to 375° and continue baking 10 to 15 minutes. Remove from oven and make a slit in the side of each puff. Return puffs to the hot, turned-off oven and leave its door ajar for 10 minutes. Cool on racks. Makes 10 to 12 large puffs, 36 to 40 small puffs. Fill with your favorite pudding or ice cream.

FROZEN STRAWBERRY DELITE

Crust and Topper Mix:
¼ c. brown sugar
1 c. stone ground whole wheat flour
1 stick soft margarine
1 c. walnuts

Filler:
1 tblsp. lemon juice
1 10-oz. box frozen strawberries, thawed
3 unbeaten egg whites
1 c. sugar
1 pint whipped topping

Crust and Topper Mix: Crumble together like pie crust and press lightly into 9 x 13-inch pan. Bake 20 minutes. Cool. Take out ⅓ of this to save for top. Crumble remaining crust and press back into pan evenly. Set aside while mixing filler.
Filler: Beat at high speed for 15 minutes. Stir in 1 pint of whipped topping. Pour over the crust and sprinkle remaining crumb topper over filling. Place in freezer at least 2 hours before serving.

FRUIT COCKTAIL PUDDING

1 c. stone ground whole wheat flour
1 c. sugar
1 tsp. soda
1/4 tsp. salt
2 tblsp. cornstarch
2 c. fruit cocktail, drained
2 eggs, beaten
1/2 c. brown sugar
1/2 c. chopped nuts
2 tsp. vanilla

Combine flour, sugar, soda, salt and cornstarch. Stir in fruit cocktail. Add beaten eggs. Pour into greased 8-inch round baking dish. Mix sugar and nuts and sprinkle over batter. Bake at 350° for 40-45 minutes. Serve warm with cream or ice cream and additional fruit cocktail.

HONEY GINGERBREAD

1/2 c. brown sugar
1/2 c. shortening
1 egg
1 c. honey
1 c. buttermilk
1 tsp. grated orange rind
1 1/4 c. all-purpose flour
1 c. stone ground whole wheat flour
1 tsp. baking powder
1 tsp. baking soda
1/2 tsp. salt
1 tsp. all-spice or cloves
1 tsp. cinnamon
1 tsp. ginger

Preheat oven to 350°. Cream brown sugar and shortening. Add egg, honey, grated orange rind and buttermilk. Beat for several minutes. Add dry ingredients that have been sifted or mixed well together. Beat until well blended. Grease and lightly flour two 8-inch square pans. Divide dough into the pans and smooth out with rubber scraper. Bake approximately 30 minutes or until top springs back when touched lightly. Place on cooling rack. Serve while warm with whipped topping or warm orange or lemon sauce. Makes 16 medium servings.

INDIVIDUAL WHEAT PUDDINGS

2 c. cooked wheat kernels
2 eggs, slightly beaten
2 tblsp. melted butter OR margarine
2 c. milk
3 tblsp. sugar
1/2 tsp. salt

Set oven temperature at 325°. Combine all ingredients; mix well. Divide among 4 to 6 baking cups. Set in pan of warm water. Bake 45 minutes or until firm in center.

LAYERED PUDDING DESSERT

¾ c. stone ground whole wheat flour
¾ c. all-purpose flour
¼ tsp. salt
¾ c. oleo
⅓ c. chopped nuts
1 8 oz. pkg. cream cheese
1 c. powdered sugar, sifted
1 9 oz. whipped topping (divided)
2 boxes instant Pistachio Pudding Mix
3 c. milk
¼ c. chopped nuts

Measure flours and salt, cut in oleo till blended. Blend in nuts. Spread and pat crumb mixture in buttered 7½ x 11½-inch baking dish. Bake in 350° oven for 25-30 minutes. Set aside to cool. 2nd Layer: Combine cream cheese with powdered sugar. Beat until blended. Add half of whipped topping and beat at low speed until blended. Spread over cooled, baked crust. 3rd Layer: Add milk to instant pudding mix. Beat about 2 minutes. Refrigerate for 10 minutes. Spread chilled pudding over second layer and refrigerate at least 15 minutes to set. Top with remaining cream cheese mixture and sprinkle with coarsely chopped nuts. (Good with lemon, butterscotch, or chocolate puddings.)

PEANUT BUTTER APPLE CRUMBLE

½ c. stone ground whole wheat flour
¾ c. sugar
¼ c. peanut butter
6 cooking apples
¼ c. sugar
½ tsp. grated orange rind
2 tblsp. orange juice
2 tblsp. water

Measure flour and ¾ c. sugar into mixing bowl. Cut in margarine and peanut butter with pastry blender or 2 knives until crumbs resemble coarse meal. Pare, core and slice apples. Arrange in 1½-quart casserole or shallow baking dish. Sprinkle with the remaining sugar, orange rind, orange juice and water. Cover with crumb mixture. Bake at 350° until apples are tender, about 45 minutes. Serve warm, topped with ice cream. Makes 6 servings.

An understanding heart is better than an understanding mind.

PINWHEEL FRUIT COBBLER

3 c. cooked fruit
Sugar, to taste
Cornstarch

Sweeten fruit to taste. Thicken juice, allowing 1 tablespoon cornstarch per cup of juice. Fold fruit into thickened juice. Pour into greased 9 x 13-inch baking pan.

Dough:
1 c. stone ground whole wheat flour
1 c. all-purpose flour
4 tsp. baking powder
¾ tsp. salt
3 tblsp. sugar
6 tblsp. shortening
1 egg, slightly beaten
¾ (scant) c. milk
Butter
½ c. brown sugar
½ c. finely chopped walnuts

Stir flours before measuring. Add baking powder, sugar and salt. Sift again. Cut in shortening until mixture resembles coarse corn meal. To beaten egg add enough milk to make ¾ cup. Add to flour-fat mixture. Stir with a fork until all of the dry mixture is moistened. Turn onto a floured board, knead lightly for about ½ minute. Roll into a rectangle about ¼ inch thick. Dot with butter, sprinkle with brown sugar and chopped nuts. Roll as for jelly roll. Cut slices of roll 1½ inches thick. Place rolled dough slices on top of thickened fruit. Bake in a 400° oven for 20 to 25 minutes. Serve warm with cream (not whipped). Serves 8 to 10. Note: Any fruit may be used in this recipe. If fresh or canned peaches are used, add 1 teaspoon almond extract.

PRETTY PIZZA

2 crust whole wheat pie dough
1 pkg. 8 oz. cream cheese
½ c. sugar
2 eggs
2 tblsp. lemon juice
1 c. each: sliced bananas, strawberries, mandarin oranges, grapes and blueberries

Glaze:
2 tblsp. cornstarch
2 tblsp. sugar
¼ tsp. mace
⅔ c. orange juice
½ c. red currant jelly

Prepare pie dough. Roll and pat into pizza pan. Crimp edges and bake at 350° for about 10 minutes. Blend cream cheese, sugar, eggs (1 at a time) and lemon juice. Pour this mixture into pie crust and bake for 10 to 12 minutes. Cool. Then arrange the fruits listed, or any variety, on top of cheese filling. Prepare glaze of cornstarch, sugar, mace and orange juice and jelly in saucepan. Cook until thick. Cool slightly, and pour over fruit.

PUMPKIN PIE SQUARES

1 c. stone ground whole wheat flour
½ c. quick-cooking rolled oats
½ c. brown sugar, firmly packed
½ c. butter
1 lb. can pumpkin
1 13½ oz. can evaporated milk
2 eggs
¾ c. sugar
½ tsp. salt
1 tsp. ground cinnamon
½ tsp. ground ginger
¼ tsp. ground cloves
¼ c. chopped pecans
½ c. brown sugar, firmly packed
2 tblsp. butter

Combine flour, rolled oats, ½ c. brown sugar and ½ c. butter in mixing bowl. Mix until crumbly, using electric mixer on low speed. Press into ungreased 13 x 9 x 2-inch pan. Bake at 350° for 15 minutes. Combine pumpkin, evaporated milk, eggs, sugar, salt and spices in mixing bowl; beat well. Pour into crust. Bake at 350° for 20 minutes. Combine pecans, ½ c. brown sugar and 2 tblsp. butter, sprinkle over pumpkin filling. Return to oven and bake 15 to 20 minutes or until filling is set. Cool in pan and cut in 2-inch squares. Makes 2 dozen.

RHUBARB SQUARES

Crunch crust:
¾ c. shortening
1 c. brown sugar
1 c. stone ground whole wheat flour
1 c. all-purpose flour
2 c. rolled oats
1 tsp. soda
1 tsp. vanilla

Filling:
4 c. cut-up rhubarb
2 c. sugar
3 tblsp. cornstarch
1 tsp. red food coloring
½ tsp. almond flavoring

For crust, melt shortening. Pour over all other crust ingredients which have been mixed together. Mix like pie crust. Spread half of the mixture in 9-inch square pan or 7 x 11-inch glass pan. Spread filling, made according to directions below, over crust, then top with the other half of the crust. Press down lightly. Bake 20 minutes at 350°. Delicious hot or cold. Serve with ice cream or whipped topping.
Filling: Cut rhubarb fine (½-inch pieces) add a little water to cook. When tender, add sugar and cornstarch mixed together, red coloring and flavoring. Stir over heat until thickened. Remove from heat and pour over bottom crust.

SAUCY APPLE DELIGHT

Batter:
1 c. stone ground whole wheat flour
2 tsp. baking powder
3/4 c. brown sugar
1 c. raisins
1/2 tsp. salt
1 tsp. vanilla
1/2 c. milk
2 large apples, peeled and shredded

Sauce:
3/4 c. brown sugar
1/4 tsp. nutmeg
1/2 tsp. cinnamon
1/4 c. butter
2 c. boiling water
2 large apples, peeled and shredded

Combine batter ingredients; blend well. Pour into a greased 2-quart baking dish. Prepare the sauce by combining sugar, nutmeg, cinnamon, butter and boiling water. Stir until butter is melted. Add apples. Pour over batter — DO NOT STIR — sauce floats on top. As the pudding bakes, the sauce seeps to bottom. Bake uncovered at 375° for 30 minutes. Serve warm, plain or with a dollop of whipped cream or ice cream.

SPECIAL BROWN BETTY

8 c. sliced, pared apples
3/4 c. apple juice
1/2 c. raisins
1/2 c. honey
1/4 c. brown sugar
3 tblsp. flour
1 tsp. cinnamon
1/2 c. oatmeal
1/2 c. stone ground whole wheat flour
1/2 c. wheat germ
1/2 c. sunflower seeds
1/4 c. honey
4 tblsp. margarine, melted

In large bowl, combine ingredients in first column. Turn into 11 x 7 x 1-inch dish. Combine dry ingredients from second column; stir in honey and melted margarine. Mix well. Spread over apple mixture and bake at 350° for 45 to 50 minutes.

Cheerfulness is contagious, but don't wait to catch it from someone else. Be a carrier.

WHEAT-APPLE CRISP

2 c. cooked wheat kernels
20 oz. can prepared apple pie filling
1 tblsp. lemon juice
1 c. brown sugar
½ tsp. cinnamon
¼ tsp. salt
¾ c. stone ground whole wheat flour
6 tblsp. butter
½ c. chopped nuts

Combine wheat, apples, lemon juice, ½ c. brown sugar, cinnamon and salt in shallow, buttered baking dish. Mix flour and remaining sugar. Cut in butter until mixture is crumbly. Stir in nuts. Sprinkle over wheat-apple mixture. Bake at 350° for 30 minutes. Serve warm topped with whipped cream. Makes 6 servings.

WHOLE WHEAT CHERRY CRISP

1 can cherry pie filling
½ c. stone ground whole wheat flour
½ c. old fashioned rolled oats
2 tblsp. toasted wheat germ
2 tblsp. bran (optional)
¼ c. sugar
⅓ c. margarine

Pour pie filling into a 9-inch round pan. Combine remaining ingredients in food processor or by hand until mixture is crumbly. Pat topping evenly and lightly over cherries. Bake in conventional oven at 350° for 35 to 40 minutes or place in microwave on high power for eight minutes, rotating once after four minutes. Serve warm or cold, plain or with vanilla ice cream or whipped topping. Serves 6. Variation: Blueberry or apple pie filling can be substituted for cherries.

WHOLE WHEAT ECLAIRS

¾ c. whole wheat flour
¼ c. butter
1 c. water
2 large eggs
Chocolate glaze or warm honey
1 tsp. instant coffee
½ tsp. vanilla extract
1-2 tblsp. plain yogurt
½ c. whipped cream cheese
1-2 tblsp. honey

Sift flour onto plate or sheet of paper. Heat oven to 425°. In medium saucepan, bring butter or margarine to boil. Remove from heat. Add all the flour to the liquid; beat well with wooden spoon until mixture forms a smooth ball that leaves the sides of the pan clean. Beat in eggs until dough is smooth and glossy. Place batter in pastry bag with ½ to ¾ inch nozzle; or place dough in plastic bag then make ½ to ¾ inch cut across one corner. Pipe from bag onto greased cookie sheets in 4-inch lengths. Bake 30 minutes in 425° oven; remove from oven and make slit in long side of each eclair. Return to the oven for 3 to 4 minutes, then lift to a rack to cool. Cool completely before filling. For filling, combine instant coffee, vanilla extract and yogurt; stir to dissolve coffee. With spoon or mixer, blend coffee mixture with cream cheese and honey, beating until light and fluffy. Spoon or pipe filling into eclairs. Makes 12 eclairs.

WHOLE WHEAT PUDDING

¾ c. cooked wheat kernels
¾ c. milk
¾ c. seedless raisins
1 egg yolk
2 tblsp. honey
1 tsp. vanilla
¼ tsp. cinnamon
⅛ tsp. salt
1 egg white

Combine cooked wheat, milk and raisins in a 2-quart saucepan. Blend together egg yolk, honey, vanilla, cinnamon and salt. Add to wheat mixture. Cook over medium heat, stirring constantly until mixture boils. Reduce heat and cook, stirring constantly until mixture is thick and creamy, about 5 minutes. Remove from heat and cool pan 5 to 10 minutes in ice water. Beat egg white until stiff peaks form. Fold into wheat mixture. Spoon into dessert dishes and chill until serving. Makes 4 servings.

The will of God will not take you where the grace of God cannot reach you.

MISCELLANEOUS

CRACKED WHEAT CEREAL

1 c. cracked wheat
2½ c. boiling water

Add cracked wheat slowly to boiling water, stirring briskly. Cover and simmer 15 to 20 minutes (7 to 10 minutes if you prefer it more chewy). Let stand a few minutes. Serve with milk and honey OR brown sugar OR white sugar to taste.

GRANOLA

7 c. rolled oats
1 c. wheat germ
1 c. stone ground whole wheat flour
1 c. wheat bran
¾ c. sunflower seeds
¼ c. sesame seeds
1 c. coconut
1 c. dry milk solids
1 c. nuts, your choice
Spices, cinnamon, nutmeg, etc., to taste
1 c. honey
1 c. oil

Combine dry ingredients in a large bowl. Combine liquids; pour over dry ingredients and mix well. Bake at 300° in large greased baking pans 30-60 minutes, stirring often. When cool, add, as desired, raisins, chopped dates, dried apples, or any fruit you prefer. Store in tightly covered container.

GRAIN AND FRUIT CEREAL

1 c. cornmeal
1 c. cracked wheat
1 6-oz. pkg. chopped mixed dried fruit
½ c. slivered almonds, toasted
½ tsp. ground cinnamon

Stir together the cornmeal, wheat, fruits, almonds and cinnamon in an airtight storage container. Cover and store at room temperature. Makes 2½ cups. To make 1 serving: in a small saucepan bring 1 cup water and a dash salt to boiling. Slowly stir in ⅓ cup of the cereal mixture. Simmer the mixture uncovered for 10 to 15 minutes or to desired consistency. Serve with sugar, OR honey, and milk.

WHOLE WHEAT CEREAL FOR BREAKFAST

2½ c. water
1 c. whole wheat kernels
Salt
Brown sugar AND molasses

Wash wheat kernels and drain. Place in a sauce pan; add water and salt and bring to a boil. Cover with lid, lower heat to simmer; cook for about 3 hours (do the evening before and reheat a portion for breakfast). This may be refrigerated and reheated as needed by using a double boiler, a microwave or heating slowly in a covered sauce pan. Sometimes a little water should be added. Variation: add raisins or chopped dates when reheating. Add brown sugar and/or molasses when serving. Serves 4 or more.

CARAWAY WHEAT WAFERS

½ c. butter OR margarine
⅓ c. sugar
¾ c. unsifted all-purpose flour
¼ c. unsifted medium rye flour
½ c. stone ground whole wheat flour
¼ tsp. baking powder
¼ tsp. salt
1 tsp. grated orange peel
2 tsp. caraway seeds, divided
2 tblsp. cold water
1 tblsp. milk

Preheat oven to 375°. Lightly grease 2 large cookie sheets, set aside. In medium mixer bowl with electric mixer at medium speed, cream butter or margarine and sugar. In separate bowl combine flours, baking powder and salt. Gradually add dry ingredients to the creamed mixture. Stir in orange peel and 1 tsp. caraway seeds. Gradually add water, mixing until moistened. Divide dough in half. On a lightly floured surface roll half the dough to ⅛-inch thickness. With a 2-inch round cookie cutter cut out dough circles. Place on cookie sheets. Brush with milk and sprinkle with remaining caraway seeds. Repeat with remaining dough, rerolling scraps. Bake for 8 to 10 minutes until very lightly browned around the edges. Makes 48 crackers.

CHEDDAR CHIPS

1 c. stone ground whole wheat flour
¼ c. wheat germ
½ tsp. salt
¼ c. butter
ice water
1 c. grated sharp cheese
¼ c. finely ground walnuts
1 egg yolk

Preheat oven to 375°. Place flour, wheat germ and salt in a bowl and mix well. Work in the butter with fingertips until mixture is like coarse oatmeal. Stir in cheese, walnuts, egg yolk and enough water to make a dough. Knead briefly. Roll out until ⅛-inch thick and cut in triangles. Place on greased sheet and bake 12-15 minutes. Cool and store in airtight container.

CHEESE STIX

2½ c. all-purpose OR bread flour
2 pkgs. dry yeast
3 tblsp. sugar
¼ tsp. garlic salt
1½ c. water
4 tblsp. oil
1 egg
8 oz. shredded cheddar cheese
3 c. stone ground whole wheat flour
1½ c. all-purpose flour

In a large bowl combine 2½ c. flour, yeast, sugar and garlic salt. Heat the water, oil and cheese to 115°. Add the warm liquid and egg to the dry mixture. Mix thoroughly and add remaining flour. Let raise about 1 hour. Punch down. Divide dough into 36 pieces. Roll each piece into an 8-inch rope. Place 1 inch apart on cookie sheet. Let raise in a warm place until doubled. Brush stix with mixture of 1 egg white slightly beaten and 1 tblsp. water. Sprinkle with poppy seed, sesame seed or coarse salt. Bake at 375°, 18-20 minutes.

CRISPY CHEESE WAFERS

2 sticks margarine
1 pkg. (10 oz.) extra-sharp cheddar cheese
¾ c. all-purpose flour
1 c. stone ground whole wheat flour
½ tsp. salt
Cayenne pepper to taste
2 c. crisp rice cereal

Place shredding disc in food processor; shred cheese. Remove disc, leaving cheese in workbowl, and insert metal knife blade. Cut both sticks of margarine into fourths and add to workbowl. Run machine until cheese and margarine are blended. Add flour, salt and Cayenne pepper; run machine just until ingredients are blended. Add crisp rice cereal and use pulse button to mix well. Roll into small balls, place on ungreased cookie sheets and flatten balls with fork. Bake in 375° oven 10 minutes, or until wafers are lightly browned. May be stored in airtight container at room termperature or in freezer. Serve warm or at room temperature. Makes about 100 wafers.

EASY WHOLE WHEAT CRACKERS

1½ c. stone ground whole wheat flour
1 tsp. salt
¼ c. oil
½ c. warm water
1 pkg. dry yeast

Mix flour and salt together. Add oil, mix well (with hands). Dissolve yeast in water. Add to flour mixture, stirring as you pour. Knead thoroughly. Let dough stand in warm place 1/2 hour. On well-oiled baking pan roll dough thin (1/16 - 1/8 inch). Cut in desired shapes. Sprinkle tops with sesame seeds. Roll lightly so seeds stick. Bake at 325° 15-20 minutes or until crisp. (If a sweeter flavor is desired, add 2 tblsp. sugar.) Options: instead of sesame seeds sprinkle tops with poppy seeds or other seasonings such as garlic salt or pizza seasonings. Makes 24-36 crackers, depending on size.

GRAHAM CRACKERS

2 c. stone ground whole wheat flour
1 c. all-purpose flour
1/4 tsp. salt
3/4 tsp. baking soda
1 tsp. baking powder
1/2 c. shortening
3/4 c. packed brown sugar
1 tsp. vanilla
1/4 c. milk

Cream shortening and sugar, then add vanilla. Sift dry ingredients and add to creamed mixture alternately with milk. Divide dough into three equal parts and roll out on floured surface 1/8-inch thick. Try to roll out into squares. Roll up loosely, or drape over rolling pin, then unroll on lightly greased cookie sheets. Use back of knife or pizza cutter to mark cracker edges in dough. Bake in oven preheated to 350° 10-12 minutes or until edges turn brown. Remove from oven and allow to cool before breaking apart.

PEANUT BUTTER CRACKERS

1 c. stone ground whole wheat flour
1 c. all-purpose flour
1 tsp. salt
1/2 c. creamy peanut butter
2 tblsp. corn oil margarine
1/2 c. skim milk (about)

In medium bowl stir together flours and salt. With pastry blender or 2 knives cut in peanut butter and margarine until fine crumbs form. Gradually add enough milk, tossing to blend well, to form a stiff dough. Press dough firmly into ball with hands. Divide into thirds. Flatten one-third dough slightly and roll very thin (less than 1/16-inch) on lightly floured pastry cloth or board. (Dough will be firm.) Cut into 2-inch squares, diamonds or rounds. Place on ungreased cookie sheet. Pierce with fork. Brush with milk and if desired, sprinkle with seeds such as sesame, caraway, celery or poppy. Repeat with remaining dough. Bake in 375° oven 5 to 8 minutes or until brown. Cool on wire racks. Store in tightly covered container. Makes about 8 dozen crackers.

WHOLE WHEAT BISCUIT CRACKERS

1 c. stone ground whole wheat flour
1 c. all-purpose flour
2 tsp. baking powder
1 tsp. soda
1/3 c. sugar
1/2 tsp. cream of tartar
1/2 tsp. salt
1/2 c. margarine
3/4 c. buttermilk

Combine dry ingredients, cut in margarine as in pie crust. Add buttermilk. Stir with fork just till dough follows fork. Roll very thin,* cut in squares as crackers. Bake 12-15 minutes in 350° oven, or bake dry as crackers.
*Use stone ground whole wheat flour to roll the dough.

PARMESAN CHEESE STICKS

1/2 c. cereal crumbs
1 1/2 oz. grated Parmesan cheese
1/4 tsp. garlic salt
Whole wheat bread
Butter OR margarine, melted

Trim crusts from bread. Cut each slice into 4 strips. Dip strips into melted butter, then roll strip in mixture of crumbs, cheese, garlic salt. Place on cookie sheet. Bake at 425° about 7 minutes. Serve with soup or salad or main course.

CROUTONS

Baked Croutons: Trim crusts from 4 slices whole wheat bread (1 cup). Butter both sides generously. Sprinkle with 1/4 tsp. garlic powder or other herb or spice. Cut in 1/2-inch cubes. Bake in 400° oven on ungreased cookie sheet, stirring occasionally, until golden brown and crisp, 10-15 minutes.

Cheese Croutons: Saute tiny cubes of day-old whole wheat bread in butter over low heat until golden brown. Sprinkle with finely grated Cheddar or Parmesan cheese and stir until each cube is coated with cheese.

TOASTY CROUTONS

3 slices stale whole wheat bread
1/4 c. butter OR margarine
Onion OR garlic salt, optional

Trim crusts from bread and cut into 1/2-inch cubes. Melt butter in skillet on medium heat. Add bread cubes; toss and stir until golden brown—about 5 minutes. Sprinkle with onion or garlic salt, if desired. Makes 1 1/2 cups. (Croutons add interest to salads, vegetables or soups. You also can use them as a casserole topping or as an accompaniment for soft-cooked eggs).

CHEESE JAM SQUARES

5 slices whole wheat bread, toasted
Butter, softened
Fruit jam
¾ c. finely shredded Cheddar cheese
½ tsp. Worcestershire sauce
1 egg white

Butter toast; trim crusts; cut each slice into 4 squares. Place on baking sheet; place about ½ teaspoon jam on each square. Mix cheese with Worcestershire sauce. Beat egg white until peaks fold over. Fold cheese into egg white. Spoon onto squares. Broil until puffed and golden brown. Garnish with cherry tomato wedge and parsley, if desired. Makes 20 squares.

EGG BAGELS

1 tblsp. dry yeast (1 pkg.)
1 c. lukewarm water
2 tsp. salt
2 tblsp. sugar
3 eggs (beaten)
2 tblsp. melted margarine OR cooking oil
4-4½ c. stone ground whole wheat flour

Egg Glaze:
1 egg (beaten) with
1 tblsp. water

Soften yeast in water. Add butter, salt, sugar and eggs. Add 3 c. flour and beat well. Add remaining flour gradually (use more if necessary) to form stiff dough that comes away from sides of bowl as you stir. Turn onto floured surface and knead until smooth and elastic (5-8 minutes). Let rise in warm place until double in size (about 1-1½ hours). Punch down and form into rectangle. Cut dough into 15 pieces and roll each piece to form rope about 7 inches long and ¾-inch thick. Shape like donut. Seal edges together with water. Let rise 20 minutes. Bring 2 qts. water to boil. Drop bagels into boiling water. Turn over after bagels rise to surface and boil 5 minutes longer. Place on greased baking sheet. Brush with egg glaze. Bake 20-25 minutes (until golden brown) at 375°. To serve, slice like a sandwich bun and spread with cream cheese. Makes 16.

Egg Bagel Variations:

Healthnut bagel—spread with whipped cream cheese; top with shredded carrot, chopped nuts, raisins, and honey.

Nutty honey bagel—spread with whipped cream cheese; top with peanut butter and honey.

Apple butter bagel—spread with whipped cream cheese; top with apple butter and chopped nuts.

Garden on a bagel—spread with whipped cream cheese; top with finely chopped radishes, celery, carrots, and your favorite salad dressing.

Bagel sea slaw—spread with whipped cream cheese; top with cole slaw and tuna chunks.

SOFT PRETZELS

2 pkg. dry yeast
1½ c. warm water
½ tsp. salt
2 c. stone ground whole wheat flour
½ c. all-purpose flour

Dissolve yeast in warm water. Add salt and flour. Knead for 5 minutes. Cover and let rise 15-20 minutes. Roll into strips 6-8 inches long. Dissolve ¼ cup soda in 1 cup water. Place each strip in this solution for 1 minute, then shape into pretzels and place on a greased cookie sheet. Sprinkle with coarse salt and bake at 350° for 20-25 minutes. Serve warm.

CARROT-WHEAT FRITTERS

¾ c. sugar
1 egg
¾ c. milk
1 tsp. vanilla
1¼ c. all-purpose flour
1¼ c. stone ground whole wheat flour
1 tblsp. baking powder
½ tsp. salt
½ tsp. nutmeg
2 medium carrots, finely shredded
1 c. light raisins
Cooking Oil
Sugar (optional)

In small bowl combine sugar and egg; beat well. Stir in milk and vanilla. In large bowl stir together all-purpose flour, whole wheat flour, baking powder, salt and nutmeg. Add flour mixture to milk mixture, stirring just to moisten. Stir in carrots and raisins. Drop batter by teaspoonfuls into hot oil (365°). Fry 6 to 8 at a time for 1 to 1½ minutes per side; drain on paper toweling. Roll warm fritters in sugar, if desired. Apples may be substituted for the carrots.

FRENCH FRIED CHEESE

Assorted natural cheeses, cut into ½-inch cubes
Beaten egg
Fine dry whole wheat bread crumbs
Cooking oil
1 tsp. salt

For soft cheeses, shape crust around soft center as much as possible. Dip in egg, then in crumbs; repeat for second layer. (Thick coating prevents cheese from leaking through.) Pour oil into fondue cooker to no more than one-half capacity. Heat on range to 375°. Add salt. Transfer cooker to fondue burner. Spear cheese with fondue fork; fry in hot oil 30 seconds. Cool slightly before eating. Note: Use soft cheeses with crust (Camambert or Brie), semihard (Bel Paese or brick) or hard (Cheddar, Edam, and Gouda) cheeses.

SNACK'N WHEAT

Drain cooked wheat in sieve and place on paper towels. Salt lightly. Allow to partially dry by placing in shallow pan and in the oven at 300° about 10 minutes. Preheat fat to 360° in small deep pan. Carefully place a small amount of wheat in hot fat and when kernels raise to surface, remove with slotted spoon and place on paper towels to absorb excess grease. To make crisper—place fried wheat kernels in pie pan and roast at 350° for approximately 8 to 10 minutes. Serve on salads or eat as a snack.

CHEESY EGG PIZZA

Crust:
1 pkg. active dry yeast
½ c. warm water
2¼ c. stone ground whole wheat flour
½ tsp. salt
1 egg
¼ c. vegetable oil

Filling:
1 8-oz. can tomato sauce
½ tsp. basil
½ tsp. oregano
½ tsp. seasoned salt
½ lb. ground beef, browned
¼ c. onion, chopped
⅓ c. mushrooms, sliced
4 hard-cooked eggs, sliced
1 medium green pepper, cut in rings
2½ c. (10 oz.) Mozzarella cheese, shredded

For crust: Dissolve yeast in warm water; let stand for 5 minutes. Add flour, salt, egg and oil; stir with a fork until mixture leaves sides of bowl. Form into a ball; knead 1 minute. Roll out on lightly floured 12-inch pizza pan; flute edges. Cover; let stand 10 minutes while preparing filling.
Filling: Combine tomato sauce and seasonings; spread on dough. Sprinkle with meat, onions and mushrooms. Arrange egg slices and green pepper rings on top. Sprinkle Mozzarella cheese over eggs and green pepper rings. Bake at 425° for 15-20 minutes. Makes 4 servings.

He that can have patience can have what he will. -Benjamin Franklin

GOURMET PIZZA

Crust:
1½ c. stone ground whole wheat flour
1 c. all-purpose flour
⅓ c. grated Parmesan cheese
2½ tsp. baking powder
1 tsp. salt
¼ c. butter
¼ c. lard
¾ c. milk
Filling:
1 c. cracked wheat
2 lb. Italian sausage
1 can (8 oz.) tomato sauce
1 tsp. oregano
1 tsp. sweet basil, crumbled
1 clove garlic, minced
4 medium tomatoes, thinly sliced
Green pepper strips
½ lb. fresh mushrooms, thickly sliced
4 c. grated Mozzarella cheese
2 tblsp. grated Parmesan cheese

To prepare crust: Combine flour, ⅓ c. Parmesan cheese, baking powder and salt. Cut in butter and lard until mixture resembles coarse meal. Gradually add milk; mix at low speed on electric mixer until mixture leaves sides of bowl. Gather dough together and press into ball. Knead dough in bowl 10 times or until smooth. Divide into half. On lightly floured surface, roll each half into 13-inch circle. Transfer to two 12-inch pizza pans, buttered and dusted with Parmesan cheese; crimp edges. Partially bake in preheated 425° oven, 9 minutes. Remove from rack to cool. To prepare filling: break sausage into bits into skillet and lightly brown, stirring occasionally. Divide cooked sausage and cracked wheat into two portions. Mix together tomato sauce, oregano, basil and garlic. Assemble each pizza as follows: Evenly distribute one-half of the sauce over the bottom, one-half of the sausage and cracked wheat and sprinkle over 1 cup Mozzarella cheese. Arrange a layer of tomato slices, pepper strips and mushroom slices on top. Over all sprinkle 1 cup Mozzarella cheese and 1 tblsp. Parmesan cheese. Bake at 425° for 20-25 minutes. Makes 2 pizzas.

Love is that quality in us that makes other people blossom out and find themselves.

PIZZA

Dough:
1 c. warm (not hot) water
1 pkg. dry yeast
1 tsp. sugar
1 tsp. salt
2 tblsp. olive OR salad oil
2 c. stone ground whole wheat flour
1½ c. sifted all-purpose flour

Tomato Mixture:
6-oz. can tomato paste
½ c. water
1 tsp. salt
1 tsp. oregano, crushed
¼ tsp basil
1 tsp. fennel seed, if available

Measure water and place in bowl. Sprinkle or crumble in the yeast. Stir until dissolved. Stir in sugar, salt and oil. Add 2 c. flour. Beat until smooth. Stir in about 1½ c. additional flour. Turn out on lightly floured board. Knead until smooth and elastic. Place in greased bowl; brush top with soft shortening. Cover and let rise in warm place, free from draft, until doubled in bulk (about 45 minutes). Mix together ingredients for tomato mixture. Set aside. When dough is doubled in bulk, punch down; divide in half. Form each half into a ball; place on greased baking sheet. Press out with palms of hand into circle about 12 inches in diameter, making edges slightly thick. On each circle of dough arrange and/or spread the following:

4 oz. mozzarella cheese, sliced about ⅛ inch thick
½ of the tomato mixture
2 tblsp. olive or salad oil
2 tblsp. grated Parmesan cheese

Bake in hot oven at 400° about 25 minutes. Serve hot. Makes two 12 inch pies.

PIZZABURGERS

1 lb. ground beef
1 c. cooked cracked wheat
⅓ c. Parmesan cheese, grated
¼ c. onion, finely chopped
¼ c. ripe olives, chopped
1 tsp. salt
Dash pepper
1 tsp. oregano, crushed
6-oz. can tomato paste
6-8 slices mozzarella cheese
6-8 whole wheat hamburger buns

Combine all ingredients except mozzarella cheese and blend together. Spread on bottom half of buns. Place in pan and broil 5 to 6 inches from heat for about 10 minutes until meat is cooked. Add cheese slices and put back in oven until cheese begins to melt. Remove from oven and place other half of bun on the sandwich.

TORTILLA SHELLS

1 c. stone-ground cornmeal
1½ c. water
3 tblsp. margarine
1¼ c. stone ground whole wheat flour
1 tsp. salt

Bring water to boil in a small saucepan. Add half the margarine. Stir in cornmeal quickly; then immediately lower heat and cover pan. Let the cornmeal cook over very low heat for 5 minutes. Stir in remaining margarine and set aside to cool. Mix flour and salt. Stir in cooled cornmeal and knead, adding water if necessary (or more flour) until a soft dough is formed. Pinch off 12 pieces and roll into 2-inch balls. Flatten each ball between palms or against a board, making a flat circle. Roll with a rolling pin to 6 or 7 inches. Keep turning the circle to keep it round, and sprinkle board and rolling pin with cornmeal as needed to prevent sticking. Cook on a hot ungreased griddle for 1½ minutes on each side, or until flecked with dark spots. Line a basket or bowl with a large cloth. Stack the tortillas in bowl and keep covered with cloth. They may be made long in advance, even a day or two before needed. Heating for a few seconds on each side makes them soft and pliable for handling again.

TACO FILLINGS:

Refried beans
Cooked Cracked Wheat
Browned ground beef
Scrambled egg
Diced fresh tomatoes, onion and shredded lettuce
Your favorite cheese, shredded

TACO SAUCE:

Heat together:
1 c. thick ketchup
¼ c. water
1-2 tbslp. chili powder
⅛ tsp. garlic powder

WHOLE WHEAT NOODLES

1 whole egg
3 egg yolks
1 tsp. salt
1 tsp. chicken or beef bouillon
4 tblsp. water
1 c. stone ground whole wheat flour
1 c. white OR stone ground whole wheat flour

Dissolve bouillon and salt in water. Beat eggs well. Add flour and water. Mix well. Divide into three or four balls, roll thin. Let dry and cut into noodles.

WHOLE WHEAT SHORTCAKE

- ¾ c. all-purpose flour
- 1 c. stone ground whole wheat flour
- ¼ c. packed brown sugar
- 3 tsp. baking powder
- ½ tsp. salt
- ½ c. vegetable shortening
- 1 beaten egg
- ⅔ c. milk
- 6 c. fruit mixed with ¼ c. sugar

Stir together dry ingredients. Cut in butter or margarine till mixture resembles coarse crumbs. Combine beaten egg and milk; add all at once to flour mixture. Stir just to moisten. Spread dough in greased 8-inch baking pan, building up edges slightly. Bake at 450° oven for 15-18 minutes. Cool in pan 10 minutes. Remove from pan and cool on wire rack. Split cake horizontally and alternate layers of cake and fruit.

'SHAKE-BAKE' FOR CHICKEN

- ¼ c. stone ground whole wheat flour
- ¼ c. unbleached flour
- ¾ tsp. salt
- ½ tsp. pepper
- ½ tsp. paprika
- ¼ c. dry whole wheat bread crumbs
- 1 tsp. herbs (thyme, basil, oregano or any mixture)
- milk or water

Mix the dry ingredients together. Moisten the chicken with milk or water. Shake the pieces of chicken, a few at a time in a paper bag. Bake in oiled shallow pan at 350° for 1 to 1¼ hours, or until chicken is tender. This is enough mix for 3 pounds of chicken. NOTE: Double or triple this recipe and store the excess in a tightly covered container for a time saver.

WHEAT SPROUTS

Wash ⅓ cup wheat kernels. Place wheat kernels in bowl and cover with enough water (approx. 1 inch) for grain to swell; cover. Let stand overnight in a cool place. Drain and rinse kernels. Divide kernels into 3 1-quart jars. Cover tops of jars with several layers of cheesecloth or nylon netting. Fasten the cheesecloth on each jar with two rubber bands or a screw-top canning-jar lid band. Place the jars on their sides in a warm, dark place (68° to 75°). Once a day rinse the sprouts by pouring lukewarm water into the jars. Swirl to moisten all the grain kernels, then pour off the water. In 3 or 4 days, the wheat should sprout. After wheat has sprouted, keep refrigerated till serving time. Serve in salads, sandwiches, soups, or breads.

INDEX

BREADS ....9 - 52

MAIN DISHES . . .51 - 63

SIDE DISHES & VEGETABLES 65-73

SALADS. . . . . .75 - 84

PIES . . . . .85 - 92

DESSERTS....133 - 145

MISCELLANEOUS ....147 - 159